The American Assembly, *Columbia University*

TECHNOLOGICAL

INNOVATION

IN THE '80s

Prentice-Hall, Inc., *Englewood Cliffs, New Jersey*   A SPECTRUM BOOK

82023

Library of Congress Cataloging in Publication Data
Main entry under title:

TECHNOLOGICAL INNOVATION IN THE '80s.

"A Spectrum Book."
1. Technological innovations—United States—Con-
gresses. I. American Assembly.
T173.8.T38 1984   338.973   84-2072
ISBN 0-13-902123-X
ISBN 0-13-902115-9 (pbk.)

This book is available at a special discount when ordered in bulk quantities. Contact Prentice-Hall, Inc., General Publishing Division, Special Sales, Englewood Cliffs, N. J. 07632.

Editorial/production supervision by Betty Neville
Cover design © 1983 by Jeannette Jacobs
Manufacturing buyer: Edward Ellis

10 9 8 7 6 5 4 3 2 1

ISBN 0-13-902123-X
ISBN 0-13-902115-9 {PBK.}

PRENTICE-HALL INTERNATIONAL, INC. *(London)*
PRENTICE-HALL OF AUSTRALIA PTY. LIMITED *(Sydney)*
PRENTICE-HALL OF CANADA, INC. *(Toronto)*
PRENTICE-HALL OF INDIA PRIVATE LIMITED *(New Delhi)*
PRENTICE-HALL OF JAPAN, INC. *(Tokyo)*
PRENTICE-HALL OF SOUTHEAST ASIA PTE. LTD. *(Singapore)*
WHITEHALL BOOKS LIMITED *(Wellington, New Zealand)*
EDITORA PRENTICE-HALL DO BRASIL LTDA. *(Rio de Janeiro)*

Table of Contents

Preface

In recent years, especially in consequence of the economic recession, there has been considerable public lament that the United States has lost its edge in the fields of scientific and technological innovation. It has been alleged that other nations, notably Japan, have overtaken us in areas which we previously dominated and that we are doomed to a future as a second-rate service society.

The efforts to remedy these putative shortcomings have involved our government, our universities, and private industry. Much concern has also been expressed about the interrelationships of these institutions in the process and the roles that each should properly play in the future of an innovative economy.

In order to seek some consensus among various interested groups and institutions about the accuracy of the allegations and about recommendations for actions to address the situation as it is authoritatively perceived, The American Assembly convened a meeting at Arden House, Harriman, New York, from November 17 to 20, 1983. Participants attended from the legislative and executive branches of the government, from industry, from the universities, from organized labor, the law, and the communications media. In preparation for that meeting, the Assembly retained Dr. James S. Coles, formerly president of Bowdoin College and of the Research Corporation, as editor and director of our undertaking. Under his editorial supervision, background papers on various aspects of the issue were prepared and read by the participants in the Arden House discussions.

In the course of their discussions, the participants achieved a substantial consensus on their findings and recommendations. They did not accept the gloomy assessments that have colored some comment on the American scene. They felt that most elements of our climate for innovation were essentially healthy. However, they did note a number of troubling factors that require careful attention if our preeminence is to be preserved. They made a number of recommendations for action, especially in the arena of primary and secondary education, designed to avert a deterioration in our national capabilities. Copies of their report, *Improving American Innovation*, can be obtained by writing directly to The American Assembly, Columbia University, New York, New York 10027.

The background papers used by the participants have been compiled into the present volume, which is published as a stimulus to further thinking and discussion about this subject among informed and concerned citizens. We hope this book will serve to provoke a broader national consensus for action to regenerate and improve those elements of our society that have always inspired a strong measure of scientific and technological innovation.

Funding for this project was provided by the Richard Lounsbery Foundation; the Research Corporation; Dow Chemical Company; Pfizer, Inc.; and E. I. DuPont De Nemours and Company. The opinions expressed in this volume are those of the individual authors and not necessarily those of the sponsors nor The American Assembly, which does not take stands on the issues it presents for public discussion.

William H. Sullivan
President
The American Assembly

Acknowledgments

The excerpts on pages 117 and 118—from David Kennedy, "War and the American Character," *The Stanford Magazine,* Spring/Summer 1975 © Stanford Alumni Association—are used by permission.

The excerpt on pages 127-128 is from Eliot Marshall, " 'Scelerosis' Blamed for Economic Stagnation," *Science,* 218, October 22, 1982, 357-58, copyright 1982 by the American Association for the Advancement of Science. It is used by permission.

Fig. 1, "The Electron Microscope" is adapted from Fig. 9 in *Technology in Retrospect And Critical Events in Science,* Vol. 1, December 15, 1968, prepared for the National Science Foundation by the Illinois Institute of Technology, Contract Number NSF C535.

The remaining figures are from National Science Foundation, *University-Industry Research Relationships: Myths, Realities and Potentials,* Fourteenth Annual Report of the National Science Board (Washington, D.C.: Government Printing Office, 1982). Their source lines are as follows: Fig. 2: NSF 81-311 and Science Indicators 1980; Figs. 3 & 4: National Science Foundation 81-311; Fig. 5: James D. Marver and Carl V. Patton, "The Correlates of Consultation: American Academics in 'The Real World,' " *Higher Education,* 5 (1976), 319-35. Used by permission of Elsevier Science Publishers, Amsterdam, Netherlands; Fig. 6: Council for Financial Aid to Education, Special Tabulation, March 1982. Used by permission of the Council; and Fig. 7: Special tabulation by the Foundation Center, May 1980.

James S. Coles

Introduction

"You can't do it, you fool! It's against the second law of thermodynamics," was Spike's immediate response to a query from Bob, a green, young local businessman trying to revive an old and faltering family business. From such an unpromising beginning was born a collaboration that, in retrospect, is almost a classic example of innovation. The result was a new, low-friction, long-wear, dry bearing material which needed no lubrication—something which the DuPont Company said couldn't be done.

Bob later recalled that "if there had been any point in his business career that marked an inchoate zero hour, yet a beginning in the right direction, it was then." Bob insisted on developing a dry bearing material. He was driven to perfect this original product that was, perhaps, not even imagined by engineers with much more technical training than he. His immediate motivation stemmed from failures in the recent past when Dixon Company had tried to use nylon in the manufacture of low-friction bearings.

JAMES S. COLES *is chairman of the executive committee of the Research Corporation (a foundation for the advancement of science) where he has been a director for twenty-four years and where he served as president for fourteen years. Before joining the Research Corporation, Dr. Coles was president of Bowdoin College. He is a director of several public companies as well as a trustee of private research organizations, including the Woods Hole Oceanographic Institution. Dr. Coles is the coauthor of a textbook,* Physical Principles of Chemistry.

(It was this attempt that provoked Spike's reference to the second law. Bob had wanted to take water—a good lubricant for nylon—from the moisture-laden atmosphere of a textile mill and deposit it on a friction-warmed bearing surface of nylon.)

From his boyhood days Bob had stubbornly insisted on being different and in his view original. His continuing idiosyncratic urge made him see life through a unique pair of lenses. This resulted in ". . . a rare overview, not so out of focus that what he undertook was preternatural, but neither was it orthodox." His childhood obstreperousness indicated early that he was to go his own way, and be his own individual person.

While Bob's baccalaureate major was English, Spike had the formal training of a physical chemist and was on the chemistry faculty of a nearby university. He had come but a short time before from three years of wartime research on underwater explosives, in which sophisticated applications of the most recent concepts in the design of experiments played a major role. This experience was most pertinent for the laboratory work he would recommend to develop the new bearing material. As Bob recalls:

> The plastic bearing material which would be developed was the result of necessity (Dixon having failed with other materials) and an outgrowth of Bob's individuality. The customers wanted it. The situation called for it. The time was ripe for it, and he believed the know-how, both mechanical and chemical, was available for it. As a result of this seemingly illogical insistence on a dry bearing, one entirely without the normal materials of lubricity such as oil or graphite, not only was a product developed for Dixon's textile industry customers, but an entirely new industry was born. For a time Dixon operated two businesses, textile machinery parts and dry plastic bearings for industry in general. The former sold to a limited market, but with the latter, the world was Bob's oyster. . . . Nevertheless it was a long, often losing, battle, profitless for several years and nearly so for several more.

With Spike's technical advice, the use of a specially designed apparatus, and an experimental design based on the concepts of R. A. Fisher, only eight months elapsed before a dry bearing material, trademarked "Rulon," was developed and marketed by Dixon Corporation. Bearings made of Rulon had a thousand times the wearability of Teflon (the principal ingredient) without significantly impairing the low coefficient of friction of the virgin Teflon.

These bearings were widely accepted by the textile industry and soon were widely used in such products as television tuners, household appliances, audio cassettes, and (with modification) nose cones of missiles. Rulon bearings, completely dry, would even run against stainless steel, a notably bad bearing mate. The company grew from but a dozen employees to several hundred, and annual sales jumped from $75,000 to $150,000 to $300,000 to $1 million in successive years, with the inevitable problems of operating capital for a rapidly growing company.

The preceding anecdotal example of a successful innovation includes all of the essential components of the innovative process within a comprehensible framework. There was the perception of a need for a new product, process, or service and the motivation to meet that need; the understanding of the fundamental laws of science; the availability of a new technology and new technologically based products; an appreciation of the research process and experimental design; a single-minded, unrelenting drive toward a steadfast goal; financing at appropriate levels and willingness to invest at significant risk; collaboration of colleagues toward a common goal; managerial skill; and, throughout all, an inclination to be different but yet accept the restraints of the laws of science, a refusal to take "no" for an answer, and a bit of the maverick.

The six chapters which follow all bear on these characteristics that generally relate to much more complex situations, such as those innovations originating in the research laboratories of large organizations, universities, and industrial corporations.

Since its beginnings, the American research university has been responsible for many innovations. Stephen Muller traces the rise of this institution from the days of von Humboldt in Germany during the early nineteenth century to its establishment in America several decades later. While new understandings of nature and new concepts in science were not long forthcoming and new scientific discoveries were soon to follow, their exploitation for the benefit of the society that supported these endeavors developed more slowly. Technology transfer from the university to a manufactured product or a new technological process did not really begin until early in the twentieth century, and only in midcentury did the value of such transfer from university laboratories generally become realized. Muller's pertinent conclusions emphasize the need for good marketing to benefit from those innovations we now are able to achieve.

As is also discussed in Muller's chapter, government has played an increasingly important role in the support of university research and, indeed, within government laboratories themselves. Donald Hornig draws from his own background in science, university research and administration, and the executive branch of government in reviewing the manner in which government has participated in technological innovation. The inclusion in the Constitution of provisions for promoting "the Progress of Science and the useful Arts" indicates that the nation's founders appreciated the potential value for the Republic of the sciences.

The early involvement of the government in technical concerns centered on exploration, geography, astronomy, and navigation. From such beginnings came the later developments in agricultural and mechanical science and government sponsored technology that led to the many alternative ways to support scientific research. In terms of government funding, the penultimate came with the development of nuclear energy; the capstone to date is undoubtedly the space program. Hornig sees a continuing government concern and participation in scientific research and technology that increasingly touches the lives of all of us.

The earliest technological innovations in America were developed almost exclusively by nonacademics—either lone individuals or those in agriculture, business, or industry. During this early period, little or none originated in academe. There are numerous examples: the cotton gin, the telegraph, the telephone, the steamboat, the internal combustion engine, the airplane, and so on. Much of this inventive genius was stimulated by what might be included in today's broad term, "market pull." We know, though, that had not men like Whitney, Fuller, Morse, Edison, Bell, Ford, and the Wright brothers had a flair for innovation and a determination to succeed despite all odds, our enjoyment of the fruits of their labors would have been long delayed.

During his career as a manager of scientific research and development within large organizations—government, nonprofit, and industrial—Robert Frosch has been concerned with a different type of innovation. His experiences in science, research, engineering, and development, as well as his knowledge of the psyche of the individuals involved with these tasks, provide insight into achievements, feasibility, motivations, and deterrents—without which good management of innovation cannot take place. He opines that we must look to better innovation management in order to take maximum advantage of our currently prolific and advancing technology.

Technology transfer from the university laboratory to public use is essential if we are to recover and benefit from our investment in academic basic research. Willard Marcy's long experience in the evaluation, patenting, licensing, and exploitation of inventions from university laboratories gives him a special insight into the manner in which this particular innovation path may be enhanced or restrained.

Typically, academic inventions are pioneer in nature; rarely, if ever, are they defensive, as are most industrial inventions. This poses its own problems. Often there is no industrial base for a new technology. This was the case for the Townes's patent on the laser. Yet countless innovations and a multitude of unforeseen industrial applications in diverse fields have since come from this single pioneer invention. Many times a new company has to be founded for the operation of the new process or the manufacture of the new product. Entrepreneurial and management skills are required, as well as risk capital. Alas, some inventions are no better than the well-established products or practices they would replace; others have so limited a market that they are left useless—"orphans," soon to be forgotten.

As director of Technology Licensing at Stanford University, Niels Reimers has not only been on the cutting edge of new technologies, but he has also brought industry and university into close, working partnerships. The skillful means by which he has solved difficult licensing problems advantageously for both industry and institution have been innovative in themselves. The management of the troika of government, industry, and university in the advancement of research and development is essential for effective research programs in large institutions. Reimers provides updates on these varied facets of the research process and its support.

Cecily Selby points out that our educational goals have not included technological understanding and the encouragement of innovative thought and practice. The talent within our youth for such thinking is there, ready to be developed, but new educational

objectives, emphases, and strategies are needed to reach the potentials of all students. As other chapters have pointed out, we
have the ability to bring conceptual innovations to the marketplace. However, how much more could we benefit if technology
and innovation were among our teaching goals? From her perspective as a scientist and an educator, Selby emphasizes an underlying
thought of some other authors herein—scientific and technological
literacy is both a right and a need of citizens who wish to live
happily and productively in an innovative and technologically
oriented society. Our current deficiencies lie not with the education of those already identified as the best and the brightest, but
rather with both the creation and maintenance of a larger and
more broadly based pool from which to draw future scientists and
engineers who are capable of innovative thinking and the development of a useful understanding of these fields in all citizens.

The recent report of the National Science Board's Commission
on Precollege Education in Mathematics, Science, and Technology
(cochaired by Selby) reviews the current status of the American
educational system with respect to the objectives stated above
and makes a number of recommendations for actions to achieve
the goals explicated in Selby's chapter. Unfortunately, among the
many other recent studies and reports by other agencies (both
government and privately funded) on precollege education, little
is said about science and technology.

A healthy redundancy is present among these chapters. At the
same time there remains a variety of viewpoints from the authors'
different backgrounds, experiences, and points of view. The
totality of their chapters produces a reasonably consistent whole.

In the editor's judgment, greater and more explicit emphasis
should be given in this country both to the transfer of technology from the university research laboratory to the market and
to the public benefit that derives therefrom. With few exceptions,
the significant technological breakthroughs or innovations of the
last half century have come from university (or "university-like")
laboratories. Thus, successful technology transfer, combined with
the most essential basic research performed in university laboratories, justifies generous financial support. Funding by government
agencies and from university endowments themselves presently
poses few problems other than questions of adequacy.

Yet the need for even more support is recognized. Universities have turned to industry for additional funding. There is no question that collaboration between the university and industry can be mutually beneficial. However, there are hazards. Premises fundamental to the form and function of universities may be compromised by certain industrial requirements. A particular university-industry relationship could influence such factors as the freedom of inquiry and publication, licensing of trade secrets, conflicts of interest, loss of objectivity, direction of effort, and the choice of fields of research. These, among numerous other hazards in these relationships, deserve careful and thoughtful attention.

In their desire for industrial support, some institutions have rushed into relationships without giving these matters due regard. Fortunately, the strongest universities have anticipated these factors in developing industrial collaboration and funding. Much good can come from these new cooperative efforts.

Steven Muller

1

Research Universities
and
Industrial Innovation in America

Ever since the mid–1970s, a belief that the future well-being of the American economy depends on a renewed national commitment to technological and/or industrial innovation has become more pronounced and widespread. Those who profess this belief usually invoke the innovative character of past American economic development and then assert that in recent years the United States has begun to lose the role of international leadership in industrial, scientific, and technological innovation. In this context the idea also is advanced that American research universities have been vital contributors to innovation in science and technology in the past, and therefore a successful recommit-

STEVEN MULLER *became the tenth president of The Johns Hopkins University in 1972, and from 1972 to 1983 also served as president of The Johns Hopkins Hospital. He serves as chairman of the board of the Federal Reserve Bank of Richmond and is a trustee of both the Committee for Economic Development and the German Marshall Fund. Dr. Muller also sits on several boards of national organizations and was the founding chairman of the National Association of Independent Colleges and Universities. A specialist in comparative government, he is the author of a textbook and numerous articles in this field.*

ment to such innovation depends essentially on leading participation by American research universities. As usual, when a majority of the public subscribes to beliefs and ideas, there is some truth in them—but no one simple truth. Some reflections on the relationship of the modern American research universities to innovation in science and technology may help to sift out reality from unwarranted assumptions and reduce some confusion.

The Foundations of the Contemporary University

It is certainly true that the contemporary major research universities are distinguished by a great emphasis on science, and increasingly on technology as well. But the extent to which these universities are the fountainhead of innovation in science and technology is at least arguable. And on the record, major research universities have not been a major—not even a significant—direct source of new products for the marketplace. The major research universities do perform research, but they remain primarily teaching institutions, and their chief role is to develop and train human talent. The vital link between the major research universities and the advancement of science and technology in the United States, therefore, can be discovered mainly in the pool of talent which the universities both harbor and produce.

Today it is difficult to remember the only very recent origin of much that is taken for granted in the contemporary American university. As of now, for example, no one would argue that the whole university is dedicated to the spirit of free inquiry. Yet the fact is that this tradition is scarcely more than a century old—precisely as old, by no coincidence, as the scientific character of the modern university. In its beginnings, the university, of course, was already committed to knowledge and truth, but the knowledge was received knowledge, and the truth revealed rather than discovered. For centuries, the university as an institution was tied inextricably to established religion and served primarily to refine and transmit established knowledge and to train human minds to function within the confines of God's word and established faith.

Thus, in the early nineteenth century, when Wilhelm von Humboldt achieved the reform of the Prussian university by insisting on freedom of teaching and learning, he had in mind a highly specific concept of freedom: freedom from religious orthodoxy.

And—as important—learning took on a second meaning beyond the original definition of absorbing all that was already known: learning began to mean inquiry as well. It is useful to note that von Humboldt's reforms were of course achieved only with the support of the Prussian government, and the statesmen of Prussia supported him explicitly because they wanted to foster their state's industrial development. The Prussian government perceived the linkage between scientific training in the universities and the application of science to and in industry, and so they sponsored the emergence of the research university. Ideas that had earlier been heresy—that truth required proof rather than faith, that knowledge could be advanced by discovery, that to question the wisdom of the past was not only legitimate but indeed necessary, and that facts were so objective that *no* known fact was sacred—were ideas now embraced within the university. Professors and their students were set free to search for the new and to seek proof for discovery.

In the United States the modern research university was not fully established until the opening of The Johns Hopkins University in 1876, with an explicitly acknowledged debt to the ideas of von Humboldt. Within a few years thereafter, graduate research programs began to sprout throughout American higher education, atop the established collegiate foundations. Even before then, however, the government of the United States had also perceived the linkage between the education of talent and national development. The Morrill Act, enacted during the Civil War in 1862, fostered the establishment of colleges specifically to educate talent in the agricultural and mechanical arts so that farming and production could spread more effectively across a whole continent. The land-grant colleges were not founded as research universities, even though they later became such, but the emphasis on the practical and its application in their founding set their professors and students free from the old rigidity of religious orthodoxy and received knowledge as well.

The devices that symbolize the industrial and technological revolution of the nineteenth and twentieth centuries—whether one thinks of the steam engine, the cotton gin, the automobile, the telephone, the telegraph, the radio, the airplane—were not developed within or by the university. Indeed, the more venerable of these devices were invented before the university as an institution had itself been transformed by science. But the application, maintenance, and continued refinement of such devices throughout the American economy depended upon a pool of trained talent which was—and is—a product of the American research university. That statement requires amplification. But before that amplification can be most effectively performed, it is necessary to observe a major second stage in the evolution of the major research university in the United States—its mobilization into national service.

Transformation by Mobilization

Until World War II, the American research university as an institution became progressively more scientific, but it did not grow hugely in size, nor did it develop significant new ties to the industrial community. The most interesting evolution of the period occurred so quietly and naturally that no one ever seems even to have remarked upon it: namely, the employment of doctors of philosophy by industry. Before the 1890s, there were, in effect, no American Ph.D.s in existence, and the degree was introduced to mark the highest level of advanced preparation for an academic career. However, well before the outbreak of World War II, industry had research departments and laboratories, and, to staff these, employed Ph.Ds and used professors of science and engineering as consultants. Thus, the high quality of the research done, for example, by American Telephone and Telegraph, General Electric, and E. I. Dupont De Nemours & Company did not depend on close relationships to one or more particular universities as such, but rather on the fact that their leading scientists were drawn from the most advanced university graduates and had the same level of training as future professors.

With the outbreak of World War II, inevitably the mobilization of the whole nation also included the universities but went far beyond the traditional call of students to the colors and the enlistment of physicians, nurses, and other specialists into service. Technology played an unprecedented role in the war effort. Not only were university specialists called to work on technologically sophisticated projects, but universities were requested to sponsor new laboratories to do research for military purposes. Nor was this a short-term effort. While the war as such ended in 1945, it was followed immediately by the so-called cold war and the Korean War; and, in fact, the period of national mobilization lasted fully for at least thirty years—until the closing of the Vietnam War. To a significant extent, mobilization still persists into what is now a fifth decade. In addition to university laboratories, new government laboratories were established in large number and variety, and more and more these too drew for research staffing on the Ph.D.s coming out of the university graduate schools.

As defense technology kept widening to include space, chemical and biological warfare, electronics, and virtually all materials, the concept of the national interest irresistibly expanded to include the whole range of science and technology within the university. Public investment by government in the growth and development of university science and technology came to be regarded as a perfectly natural—indispensable—ingredient of national security. First millions, soon billions, of dollars annually were appropriated for this purpose. At the same time, access to higher education was being expanded by means of a succession of congressional enactments and appropriations. As a result, existing colleges and universities grew greatly in size, and new colleges and universities were established. In the quarter century between 1945 and 1970, American higher education more than tripled in size and capacity, and within the major research universities the federal government became the established patron of advanced research and training over the entire range of fields in science and technology.

The Government-University Partnership in Research

Selected aspects of the way things were done in the process, or of the way in which matters turned out, appear worthy of comment. For example, it can be noted that the interaction between representatives of government and the university community began in the 1940s on an extraordinarily high level of mutual trust and commonality of purpose. World War II was—at least after Pearl Harbor—a "popular" war in the United States. Subsequently there was widespread consensus that the best way to counter Stalinist expansionism and avoid renewed global war and the use of nuclear weapons was to create effective deterrent capacity. Cooperation in the national interest was not then controversial. In other words, motives were not initially in question. As a result, problems that might otherwise have led to long and vexed negotiations were settled quickly and effectively in order to get the job done. An enduring network of personal connections between individuals in government and those in university science grew in this agreeable climate, and those helped to lubricate relationships later when some friction began to develop.

It must be assumed that the high degree of mutual trust at the outset had much to do with the easy adoption of the peer review systems in the distribution of increasingly vast amounts of government sponsored research. There is, in retrospect, a near miraculous purity in the concept that the best way to assure the funding of good science is to allow good scientists to review applications and select the best. And—most of all—it is worth noting that it was possible for government to deal directly with university scientists and technology experts themselves, with only relatively minor involvement on the part of the universities or institutions. It is more than doubtful whether university administrations could have motivated professors to cooperate with government nearly as effectively as was in fact the case, where the motivation arose within the professional initiative combined with the appeal of the national interest that largely swept the institutional university along in its train.

In the well-known story of the growth of government sponsored university research, the involvement of industry is seldom mentioned or emphasized. While this may be easily explained because industry involvement was indirect, it is a grave distortion not to recognize explicitly the major stake on the part of industry in the burgeoning government-university partnership. Even if one were to look only at national defense in a narrow sense, it is obvious that the ever more sophisticated and complex national defense systems—developed with the advice of university specialists—called for an ever greater range of sophisticated and complex products—products procured by government and produced by industry. The wider the range of government needs—beyond weapons systems and into, for example, space and communications technology—the greater became the involvement of diverse industrial enterprises in providing the means growing out of research and development. It is, of course, true that in response to the situation, more and more of the affected industries began to set up elaborate research and development programs of their own—also often with government assistance. But here too the staffing of these industrial research and development programs depended on the availability of university trained talent—talent at the core, trained at the doctoral level. The great investment on the part of the federal government in university science and technology, therefore, produced not only ideas and techniques that resulted in industrial contracts, but also—and with far greater total impact—provided the funds and facilities within universities to train great new numbers of highly advanced technologists and specialists, who found employment in industry and government, as well as within the university system itself.

To the extent, however, that the federal government was not only the principal sponsor of science and technology in the major research universities, but also the principal consumer of so much of the applicable result, it can be remarked that the need to *market* ideas and techniques was generally—and notably—absent. To a large extent, government was willing to sponsor basic research, i.e., the conduct of scientific inquiry for its own sake and where an applicable outcome was neither promised nor expected. However, where the government sponsored targeted research, the government was also likely to be the consumer or purchaser of the result; hence there could also be a certain indifference as to whether the result was ever purchased or consumed—that decision was, after all, up to government. There was *competition*—among investigators for research support and among industrial enterprises for procurement contracts, but there was very little marketing.

University Attitudes toward Research

In this connection it should also be pointed out that research as a *product* is not—or, at least, not yet—an accepted notion even within the contemporary American research university. To understand this, it is useful once again to go back to Wilhelm von Humboldt and the germinal reform of the Prussian university which he achieved. Von Humboldt spoke not only of freedom of teaching and learning, but also of the identity of research and teaching. His credo was that inquiry was an indispensable part of teaching: only someone engaged in inquiry was best qualified to teach, and learning involved engagement in inquiry as much as absorption of subject matter. The twin identity of research and teaching has since become—and remains—gospel within the American research university. And this twinning needs to be understood in light of the fact that the university has been—and remains—primarily a research institution. Research without teaching is still versity styles itself as a research university, what is meant is that its teaching mission is distinguished by a research component of the highest quality. What is not meant is that the university is primarily a research institution. Research without teaching is still as heretical an idea within the contemporary American university as teaching without research.

To understand this confluence of teaching and research within the university supports the notion that the university as an institution is generally ill-suited to perform research: it is the professor *at* a university who performs research, not the institution. The key relationship which evolved as government became so prominent a sponsor of research was—as noted earlier—between government and individual professors identified as principal investigators. The inner logic of this arrangement lies in the linkage of research and teaching as well as in the freedom of inquiry: only the researcher/teacher could appropriately determine the proper mixture of inquiry and instruction that is inevitably a cardinal feature of an academic research project. Thus, on the face of it, a particular university can be identified as "doing" on the order of $100 million annually of federally sponsored research, and it is accountable to government and the public for the whole of it. But in reality so great a total is merely the accumulation of hundreds of individual projects, solicited and executed under the guidance of principal investigators, normally unrelated to each other, and scattered throughout the university. A major research university is one whose faculty is composed of many persons of such distinction so as to be able to bid successfully for research awards—grants and contracts awarded by government in the name of the institution but awarded in fact to the principal investigators. Universities did not and do not *assign* research to members of the faculty any more than they assign the courses to be taught. Instead, professors select the research they wish to do just as they select the content of their teaching, and, if funded, they thereby put the university into that particular research activity. When professors—principal investigators—move from one university to another, their research awards follow them and do not remain at the university of origin. As a result, a university widely known for research of a particular kind could—and does—suffer loss of competence with the departure of a principal investigator, whose arrival at a different university would then lend to it the distinction lost by the institution from which the move originated.

There were—and continue to be—some exceptions to this prevailing situation in that some universities did set up special laboratories, dedicated to particular lines of inquiry, which sought and received support as such, i.e., not on the basis of individual grants and contracts. In most instances, however, there was a controlling reason for such action by the university: the need for secrecy. When government insisted on secrecy in the national interest, the university faced—and still faces—a dilemma. On the one hand, it is obvious that certain types of research involving national security require the protection of secrecy lest they aid foe as well as friend; on the other hand, secret research is anathema to academic practice. Precisely because of the fundamental credo that research and teaching are inextricably linked, research that—for reasons of secrecy—cannot be related to instruction is academically illegitimate. Academic research *must* serve—or at least be capable of serving—as a teaching base and, therefore, *must* be open. By definition, then, secret research cannot be academic research. To resolve this dilemma, universities willing to engage in secret (classified) research set up nonacademic laboratories, physically isolated from the rest of the campus, in which secrecy could be maintained—but at the sacrifice of the academic mission. At the same time a decision was reached that individual faculty members could engage in secret research as a matter of individual choice, but *not* on the campus. Professors can, in other words, serve as consultants on secret or classified projects, but only if the work they did on such a basis was located outside the academic campus and as long as their laboratories and offices on campus remain entirely open. This mode of operation made it possible to achieve some academic linkage between an off-campus secret research project sponsored by a university and the same university's academic departments. By means of joint appointments, an investigator primarily engaged in secret research can come onto the academic campus as a part-time faculty member, at least to teach, but possibly to perform nonclassified research as well; a regular faculty member can leave the academic campus and engage in secret research at the classified project site, serving as a part-time consultant.

Corrosion in the Government-University Research Partnership

In the course of the 1970s, a gradual sea change occurred in the relationship between the federal government and the major research universities—a sea change hard to define both because it took place gradually and because so much on the surface remains the same, but also it was sufficiently severe so that, in effect, it seems to mark the end of an era. A series of circumstances coincided to produce this effect. First, the constant dollar level of federal government appropriation to support university research in science and technology ceased to rise, and on occasion had even fallen, not only from one year to the next but over several years in succession. A form of the cold war continued; the nation's investment in national security remained extremely high; even the countrywide mobilization in the national defense remained a constant of sorts. But as far as the universities were concerned, the context of federal research support changed from growth and renewal to contraction. And this came about in combination with the end of that earlier sustained period of growth in student and faculty numbers. Overall, most of the level of effort reached in the past still continues, but the steady acceleration of support of the previous twenty-five years has halted.

Of greater importance may be the fact that substantial corrosion has appeared in the process of government-university research interaction. This is not surprising in that it is only a natural occurrence when a relationship goes on for so protracted a period, but an understanding of this reality merely explains problems without attenuating them. To a significant degree, the initial trust and shared common purpose between government and the university community have been substantially dissipated for all sorts of reasons. The unpopularity of the Vietnam War produced sharp differences between government and the majority of the academic community. The sheer volume of federally sponsored research became so great that inevitable problems appeared in the auditing and accountability for so huge and diverse an annual effort. With the enormous growth of the professoriate over a quarter of a century came some dilution in quality. Where, early on, a relatively small elite of faculty members at relatively few institutions had dominated the interaction on the university side, there were now much larger numbers of persons from many more institutions involved, and the quality of peer group evaluation became somewhat arguable in the process. Over time, just enough instances of poor fiscal management and/or questionable performance occurred to corrode some degree of faith and confidence. And, after all, a process dependent on annual appropriations from so highly political a body as Congress could not expect indefinitely to remain miraculously untainted by political consideration. Additional corrosion therefore occurred when, on occasion, Congress began to tie strings and ribbons to federal grants and contracts. Recently, there has also been a tendency—still unchecked—to make some awards on political grounds by simply and blatantly operating outside the regular process of research proposals and peer-group review.

Other considerations entered the picture as well. Quite apart from inflation, the absolute cost of pursuing research has become steadily greater as the technology of research itself became ever more complex. The scientist doing equations on a blackboard— as fixed symbolically in the public eye by the ineradicable image of Einstein in his study—has been superseded by the research team operating with a vast laboratory array of instruments whose cost and complexity are awesome. And furthermore, the range of science and technology far outstrips even the most all-inclusive definition of national security, and the result is that real argument is now possible as to the priority for research support when weighed against the whole array of other public priorities.

The Industry-University Partnership

In the wake of this major revision in the relationship between government as principal sponsor of research in science and technology and major research universities, a still increasing effort developed to establish a new level of direct partnership in research between the university and private industry. No one has ever suggested that private industry should eventually replace the federal government as principal research sponsor; nor has it been assumed that federal research sponsorship would cease. But the assumption that federal research support in constant dollars would at best level off and perhaps also be less comprehensive has led the university to be interested in industrial research sponsorship as a supplement to—not substitute for—federal support. As for industry, the trigger comes in the field of biotechnology and genetic engineering, whose results in many instances have greater promise for commercial rather than national security development. (However, this interest may well be of limited duration. Recognizing the potential in these fields for the production of pharmaceuticals, foods, and chemicals and, initially, the almost total absence of in-house expertise in industrial laboratories, industry turned to universities and their faculties for knowledge and expertise. As in-house expertise is hired or developed, this dependence on outside university expertise will diminish and may, within a few years, be of only minor importance.) By the beginning of the 1980s, therefore, discussion among representatives of universities and private industry began to be intensive and continuous. A number of large industrial commitments for sponsorship of university research received national publicity, accompanied by a host of smaller scale, less well publicized commitments of great diversity. It appears extremely likely that direct university-industry partnerships in research will continue to proliferate. However, this new linkage has significant limitations and problems. Some have already been widely discussed; others, less so. An interesting and useful way to appraise them may take the route of comparison with the process of research sponsorship by the federal government.

Partnership grows out of mutual interest. And as noted, the foundation of the partnership between government and the universities lay in shared devotion to the national interest—specifically to national security in time of war. The analogous shared concern between industry and the universities appears to revolve substantially around financial gain: most fundamentally, profit for industry, research support for the university. How sound is that analogy? It can be argued well that financial gain represents at least as much of a mutual incentive as patriotism or even that gain can exceed patriotism in intensity. However, it may be more difficult to argue that financial gain as motive can parallel patri-

otism in serving as the basis for mutual trust. That, in turn, may
be particularly relevant to the potential of industry-university
relationships because of the dichotomy involved in university participation. As was and is true in the government-university
partnership, the operative university partners are the researchers
themselves—the principal investigators. In the partnership with
government, the basic assumption was not only that everyone
within the university shared a common commitment to the national interest, but—especially at the outset—financial gain beyond
the mere generation of support for research scarcely was perceived
as a factor; the concept of profit did not usually enter into
consideration.

In the partnership with industry, however, profit does enter into
consideration, either actually or potentially. On the one hand, it
would be unfair if a corporation made large profits from an application of university based research and there were no sharing
whatever with the university partner. On the other hand, insofar
as the university partner is both the individual researcher and the
institution, how is profit shared between these two? At first blush
one might think that this question is easily answered by drawing
on a long history of institutional patent policies that represent
both a tradition and experiential base for profit sharing on the
part of the industry as well as for profit sharing between principal
investigator and university institution. But in practice there is
the complexity involved, for example, in stock ownership by
professors and/or universities as institution and in profit sharing
by corporations with scientists who serve only as consultants on
an individual basis and not as participants in a university sponsored relationship.

It is not relevant at present to explore this and other complex
entanglements further; however, it should be noted that a great
degree of mutual trust is more apt to develop and be sustained
over protracted periods by the generation of common concern
based on patriotism than by those based on financial gain. In fact,
without excessive cynicism one must note the effort to evolve
a common industry-university concern much more analogous to
wartime devotion to national security, as at least a complement
to the profit motive. The common concern invoked in this view
is technology transfer—a phrase that stands for the common
humanitarian impulse to strive to make the benefits of applied

research available to the public as rapidly and effectively as possible. (This was the impulse governing Professor Frederick Gardner Cottrell when he established Research Corporation as a nonprofit technology transfer agency in 1912.) More recently this concept has also been directly related to the national security—by referring precisely to the discussion of economic innovation with which this chapter began. Patriotism, as well as profit, can be invoked by the argument that the welfare—and security—of the United States depends on sufficient technology transfer directly from university to industry in order to assure not only that discovery results in new benefits to the quality of human life in the best and quickest manner, but also to assure that American industry thereby remains so consistently innovative as to reclaim and retain world leadership.

The profit potential in this context then becomes a desirable but secondary enhancement of a more noble primary goal. And even those who might be reminded—skeptically—of the now famous old assertion that "what's good for General Motors is good for the country" may find it difficult to deny that there is truth in the argument that university research relates positively to innovation in industry. Obviously, however, any argument linking industrially sponsored university research to American national purpose is awkward to justify when the sponsoring corporation is a major multinational enterprise based abroad. And the fact is, of course, that at least a few of the most prominent new linkages between particular industrial corporations and American research universities have involved foreign, rather than American, enterprises.

POTENTIAL HAZARDS

There are other problems that emerge when industry-university research relationships are compared with the government-university research partnership, particularly those relating to the absence of overriding national interest as basic justification. On occasion, for example, industry would like to impose secrecy on research, but for proprietary purposes rather than by reason of national security. Universities, committed (as already indicated) to the inseparability of teaching and research, cannot appropriately accommodate industrial interest in confidentiality any more than government interest in secrecy. Ideally, therefore, confidential industrial research should be carried on by professors only off-campus, in industrial laboratories, just as was and is done with secret government research. But the presence of the profit motive makes the easy parallel more difficult to apply. What happens, for instance, when the principal investigator is also the entrepreneur? What happens when the university as an institution stands to profit through a contractual arrangement or as an investor? Are patents the answer? It is generally assumed (most conveniently) that the time required to obtain a patent is just about as long as that required for the publication of a piece of research. But will this result in an erosion of time-consuming testing because of a rush to publish? And what happens if the research in question involves unpatentable techniques that are best protected as trade secrets?

Questions such as these raise the more fundamental issue of whether the anticipation of financial gain will tend inevitably to draw professors away from the concept of research as pure inquiry toward the goal of research for profit. Earlier, goal oriented research had become something of an issue in the course of the government-university research relationship. Often, however, because the goal was classified, the research took place away from the university in any event. In the case of other goals, such as "the war on cancer," the goal was so broad and humane as to cause no problem.

Financial gain is more suspect, particularly because the university as an institution is as directly involved as the principal investigator. In the case of government sponsored research, it is assumed that the university as an institution has only a minor interest in the substance of any particular piece of research being done as long as it is not secret and as long as the principal investigator who solicited support is appropriately funded and committed. But will university administrators, representing the interests of the university as an institution, remain in such a position of benign indifference when there is a prospect of financial gain for the institution? Will there, in other words, be a tendency by the university to push professors not merely to perform research and obtain support for doing so as has long been the case, but to perform particular kinds of research with financial gain in mind?

This line of inquiry compels recognition of another relevant difference between the government-university and industry-university relationships. As noted earlier, the essence of the government-university relationship was government sponsorship through a process of open application by principal investigators whose applications were subject to a peer review process. Relationships between industrial corporations and universities increasingly have taken on an entirely different form. First, the diversity of industry and of the professoriate is so great that some sort of brokerage was required to match potential sponsors and investigators; university administrations began to play the role of broker. Second, a marketing approach emerged—a corporation marketing its interest in sponsorship and a university marketing its interest in receiving sponsorship. Third, instead of a nationwide application process and competition by application, corporate research sponsorship with a university tends to be negotiated on a one-on-one basis, and in most cases it contains no form of peer review. Fourth and finally, the result for the universities was a new and highly competitive race for industrial sponsorship in which university administrators were actively marketing the skills of their professors. It is against this background that questions are asked as to whether or not the university as an institution will attempt to impart guidance to principal investigators when the factor of financial gain is present.

Fundamental Issues for Industry-Government-University Research Interactions

Problems of this kind are significant and awkward, and they continue to be both explored in practice and debated in the abstract. They are, however, dwarfed by two other considerations that may be even more fundamental and as yet have received very little discussion. The first of these derives from further consideration of the enormous cost of research instrumentation in the universities. As noted earlier, the aggregate sum required for adequate instrumentation already appears to be growing beyond even the capacity of the federal government to sustain at public expense. And large as the collective resources of private industry may be, they fall short of the resources available to government; and in fact, no way exists (nor is likely to be found) for a *collective* application of industrial resources to support research in the universities. Even industry-by-industry collective collaboration is hamstrung by antitrust legislation; company-by-company approaches are the rule. At the moment, such approaches appear to be feasible only as long as the application of corporate resources remains a marginal supplement to a much larger volume of support from government. It follows, however, that the significant future decreases in government research support are not likely to be offset by a sufficient increase in support from industrial corporations. Instead of industrial resources rising to balance out shrinking federal allocations, a more likely prospect would be that major reductions in federal support for instrumentation and its installation and maintenance would make university laboratories *less* attractive to corporations because, rather than complementing government support, available corporate resources would become submarginal under these circumstances.

The future requirements of support for instrumentation have practical consequences for the universities, for industry, and for government. For the universities, assuming that the twin pressures of need and practical possibility will, over the long run, impose their own logic, the most likely answer would appear to lie in the type of sharing that has already evolved in the field of high-energy physics. Just as, for example, only a finite number of nuclear accelerators exists and just as these are governed by consortia of institutions so as to provide access to investigators across the entire discipline, so it seems probable that truly large-scale instrumentation resources in other scientific and technical fields will evolve along similar lines. The university or universities in conjunction with which such resources are located will, on the one hand, develop special strength in the relevant particular area of inquiry; on the other hand, colleagues from the rest of the university world will also have access to the facility and its resources.

And, at a different level, universities will need to consider more effective sharing of resources with colleges that operate on the undergraduate level only. The issue in this respect is not research —professors located in colleges will already have access for research purposes to highly advanced instrumentation resources at major research universities—but teaching. The universities draw on the collegiate sector for their graduate and professional students. Universities thus have an interest in preventing the decline of instrumentation in undergraduate colleges to the point where college graduates would be so underprepared in science and technology as to be dysfunctional in graduate and professional schools. As a result, universities will see the need to share the most expensive and sophisticated instrumentation with colleges for teaching purposes. There are new lines of sharing within higher education that as yet have barely begun to appear.

As for private industry, corporations dependent on science and technology have an unavoidable stake in the adequacy of instrumentation and the quality of research in the major research universities. The *essential* linkage between the universities and industrial innovation and vitality consists of people-related as opposed to product-related research. The article of faith within the university community which insists on the inseparability of research and teaching is not merely sacrosanct—it is practical wisdom as well. To the extent that its consequence puts limits on the direct applicability of university research precisely because that research must also serve a teaching mission, those limits are an asset rather than a liability. Both government and industry are inescapably dependent on a flow of talent which the universities produce. To a large degree, the quality of government and industry in the age of technology is determined by the quality of available talent; the stream of the most highly trained, specialized, and scientifically and technologically advanced talent flows out of the university pool.

Industry recognized long ago that innovation in science and technology depended on the creation of industrial laboratories. These laboratories, rather than university laboratories, are the proper and best source of product development. But industrial laboratories are staffed by university graduates. Under ideal circumstances, universities are the source of graduates trained in the methods of inquiry with state-of-the-art instrumentation, who are eligible to be hired by industry for its laboratories. Technology transfer occurs as well informed and highly skilled human talent moves constantly out of teaching laboratories into applied research laboratories. Nor is this a one-way street. New techniques and results from industrial laboratories move over into the teaching laboratories of universities, not only by the maintenance of personal contacts, but also because university scientists already consult sufficiently with industry to stay current with industrial

advances in science and technology. Under less than ideal circumstances, universities would lack the resources for the adequate and
most up-to-date preparation of graduates in science and technology;
the pool of the most highly trained talent would then be not fresh
but stale. At the worst, industry would itself have to offer the
ultimate in advanced training if industrial laboratories alone
were to offer advanced instrumentation no longer available in
university laboratories for research and teaching.

If this is the correct perspective, then at least a good deal of
the prevailing industrial and public fixation on the *substance* of
university research appears to put the accent on the wrong syllable.
The most fruitful outcome of the now protracted experience of
government sponsorship of university research has been in fact
the splendid enhancement of the nation's pool of most highly developed talent, not the research results obtained in any single
instance or in aggregate. Clear recognition that universities exist
to teach and that the contemporary university must do research
in order to teach—not do research for its own sake—provides the
best guidance for future courses of action. Such recognition
implies that government, industry, and the universities fully
share a common purpose: to assure the ability of the universities
to attract, nurture, and prepare human talent at the most advanced
level of science and technology so that the goals of government
and industry will not be impeded for lack of human resources.
Industry, therefore, should move beyond the current emphasis on
the possibility of product development directly from university
laboratories to a more fundamental emphasis on the preservation
and enhancement of the teaching mission of the university. In
practical terms this would mean supportive concern by industry
with the continuation of public investment by government in the
strength and quality of university research—and hence, teaching
—in science and technology and less effort on the part of corporations to leverage the prospect of financial gain for universities into
pressure to unhinge university research from the teaching mission
so as to move it closer to a more goal oriented character with
directly applicable results in view.

The future of innovation in the American economy does indeed
depend on the American university. The dependence, however,
rests far less on the results of university research per se than on
the indispensability of research to the training mission of the

university. The university's role in the development of human talent transcends by far the university's role in discovery—or, explicitly, the goal oriented quest for discovery. It follows, then, that the future of both national security and national prosperity depends significantly on a continued investment in university science and technology, supported by both government and industry, with the primary emphasis on the development of human talent.

American Shortcomings: Innovation, Productivity, or Marketing?

Tempting as it is to end here, some brief concluding observations may be useful on the innovative character of American society in comparison to other national societies. On the one hand, there *is* evidence that the United States has no monopoly among the countries of the world on innovation in science and technology. On the other hand, there is no evidence that to date the United States lacks the ideas or the talent to retain world leadership in the advancement of science and technology, provided that adequate resources are supplied. The record of recent international economic experience shows little evidence that other national economies are more innovative than that of the United States insofar as the substance of science and technology is concerned. What that record does reveal, however, is that other nations have been and remain more innovative and successful in production, manufacture techniques, and international marketing of new products than the United States. Japan, for example, in addition to superior production and quality control, appears to be applying a genius for the identification and exploitation of world markets through a combination of highly innovative and effective product development and marketing far more than striving for original discovery in science. In contrast, American industry continues to draw on original discovery but may be falling behind in international market share. For this there may be several reasons. First, American corporations may be too comfortable with a domestic market of continental size which for decades has been familiar, as well as sufficient to sustain profits and growth. Second, American corporations have relied heavily on foreign employees when selling abroad, not only on the assumption that indigenous citizens of other lands will get the best reception within their domestic markets, but also because American talent familiar in depth with foreign markets is in extremely short supply. Third, there has been a reluctance to invest in new manufacturing facilities and an inability to control quality of product.

The simple fact appears to be that aside from innovation in the manufacturing process, American corporations need to cultivate foreign markets more effectively on a global scale and, in the process, rely more on American talent that knows the area, its culture and history, and, above all, its languages. To the extent that this is true, American universities may have a major contribution to make to national economic prosperity, not only through teaching and research in science and technology, but also through foreign language and area studies for far greater numbers of students than have participated in the past. It is not true that worldwide marketing is a new concept for American industry, but it may be true that worldwide marketing falls short when it is executed and supervised by Americans who speak only English and on behalf of products designed primarily for an American market. The major research universities in America are among the most cosmopolitan, least parochial institutions in the country. Their ability to provide human talent familiar in depth with any and all areas of the world may need greater recognition and support in the context of national prosperity within a global economy. The earlier research partnership between government and the universities included language and area study centers and fellowships, long and successfully supported by the National Defense Education Act. It is worth considering whether American industry has a major stake in reviving and supplementing that experience as well.

In summary, then, the power and strength of American industry in a global economy depends both on future innovation *and* the capacity to market the results worldwide. Innovation in this era of science and technology depends on numerous factors—one of which without doubt is human talent of appropriate high quality. The American university has become the proper training ground for such talent by virtue of the effective linkage of scientific research to its traditional teaching mission. Thus, industry and government have a joint stake in university research, less for the sake of applicable results than for its indispensable educational function. And if there is truth in the thought that American industry may be more deficient in marketing than in innovation, the universities have the capacity to contribute to the solution of that problem as well.

2

The Role of Government
in Scientific Innovation

Introduction

The notion that innovation is a virtue to be fostered in its own right is a very recent one. In fact, the whole idea of progress as something that can be systematically fostered is largely a twentieth century phenomenon. To be sure, the ideal of systematically acquiring knowledge and using what is learned to develop an understanding of the physical world goes back 500 years or more, but the current concern with innovation goes much further. By scientific innovation we mean the process by which new knowledge and skills are generated and applied to the social, economic, and intellectual operations of society. Innovation, then, is more than discovery and theorizing, more than speculation and invention, and more than engineering design. For until the

DONALD F. HORNIG *is the Alfred North Whitehead Professor of Chemistry at the Harvard School of Public Health. Previously, Dr. Hornig served as director of the Office of Science and Technology and was special assistant to the President during the Johnson administration. He joined Eastman Kodak Company as vice president and director before becoming president of Brown University. Dr. Hornig is active on committees of numerous national institutions, is a noted lecturer, has addressed international congresses and learned societies, and has published over eighty-five scientific articles.*

new "know-how" is incorporated into what is done by our economy and our society, innovation in our sense has not occurred.

The problem to which this chapter is addressed is what the role of government can be in encouraging scientific and technological innovation. Other governments have gone beyond the encouragement of innovation to attempts to guide the innovative process or even to carry it on in government agencies or laboratories. Our concerns are whether or not the U.S. can remain technologically competitive in the face of the more or less centralized technology policies of Japan, West Germany, and France and what role the government can play. We shall look at our own experience to date and later explore possibilities which are available, some of which have been tried elsewhere.

BENJAMIN FRANKLIN

Before proceeding it is interesting to remind ourselves that while fostering innovation may be predominantly a post–World War II phenomenon, it actually has roots in the eighteenth century. In 1743 Benjamin Franklin circulated "A Proposal for Promoting Useful Knowledge. . . ." With a foresight and vision which is rare in developing countries and often in short supply in the advanced industrialized countries, he wrote:

> The English are possessed of a long Tract of Continent, extending . . . thro' different Climates, having different Soils, producing different Plants, Mines and Minerals, and capable of different Improvements, Manufactures, etc.

> The first drudgery of Settling new Colonies . . . is now pretty well over; and there are many . . . in Circumstances . . . that afford leisure to . . . improve the common stock of Knowledge. To such of these who are Men of Speculation, many Hints must from time to time arise, many Observations occur, which if well-examined, pursued and improved, might produce Discoveries to the Advantage of . . . the British Plantations, or to the Benefit of Mankind in general.

He goes on to propose a "Society of ingenious men" to maintain a constant correspondence concerning:

> All new-discovered Plants, Herbs, Trees, Roots, etc. and their Virtues, Uses, etc.; Methods of Propagating them, and making such as are useful but particular to some Plantations, more general; Improvements of vegetable Juices, as Cyders, Wines, etc.; New Methods of Curing or Preventing

Diseases; All new-discovered Fossils in different Countries, as Mines, Minerals, Quarries; etc. New and useful Improvements in any Branch of Mathematicks, New Discoveries in Chemistry, such as Improvements in Distillation, Brewing, Assaying of Ores; etc. New Mechanical Inventions for saving Labour; as Mills, Carriages, etc.; and for Raising and Conveying of Water, Draining of Meadows, etc.; All new Arts, Trades, Manufactures, etc. that may be proposed or thought of; Surveys, Maps and Charts of particular Parts of the Sea-coasts, or Inland Countries; Course and Junction of Rivers and great Roads, Situation of Lakes and Mountains, Nature of the Soil and Productions; etc.; New Methods of Improving the Breed of useful Animals; Introducing other Sorts from foreign Countries. New Improvements in Planting, Gardening, Clearing Land, etc.; And all philosophical Experiments that let Light into the Nature of Things, tend to increase the Power of Man over Matter, and multiply the Conveniencies or Pleasures of Life.

This proposal led to the establishment of the American Philosophical Society. But it is also an agenda for innovation which in spirit could guide us today, even if the details need some updating. Of course, correspondence is no longer enough. Our question is, what more needs to be done to achieve such goals in the twentieth century?

THE EARLY REPUBLIC

Despite these ideas, science did not find its way into the Constitution of 1789, even though Franklin and Jefferson participated in writing it. Along with universities, canals, a central bank, and other such projects, scientific projects were thought of as "internal improvements," public works which might lead to centralization of power in the federal government and which were therefore best left to the states. The Constitution of 1789 mentions science only once, in the power of Congress to "promote the Progress of Science and useful Arts, by securing for limited Times to Authors and Inventors the exclusive Right to their respective Writings and Discoveries." Thus was patent and copyright protection born.

President Thomas Jefferson hoped for more. His love for science was such that he had a room in the new executive mansion in which he worked on fossil bones, one of many scientific projects he undertook before, during, and after his Presidency. To involve the government, he proposed in 1806 that "public education, roads, rivers, canals, and such other objects of public improvement as it may be thought proper to add to the constitutional enumeration of Federal powers" be carried out with federal funds, and he sought a constitutional amendment to that end. His favorite project was a national university which would conduct both "research and instruction." None of this came to fruition since the constitutional amendment failed.

Jefferson did succeed in two important projects. In 1803 the Lewis and Clark expedition to the Pacific Northwest explored the continent and made significant findings in botany, zoology, and ethnology. As well as being successful in itself, the Lewis and Clark expedition paved the way for a long period of exploration of the continent, the Arctic, and the surrounding seas—supported by the federal government. Related to the explorations was the organization of the Coast Survey in 1807. Better charts, navigational aids, and topographical information were henceforth to become a federal responsibility. An era of exploration and surveys was inaugurated which continues to the present day in the Coast and Geodetic Survey, the explorations of Antarctica, and the mapping of the solar system.

Not mentioned in these generalizations is how continual political opposition to the involvement of the federal government in "internal improvements" led all of these ventures (and others) to proceed in fits and starts. The institutions to carry science forward evolved slowly from these fragments, and the most durable organizations were the army and navy. That is a lesson which is still with us—that it is easier in many cases to undertake new ventures under the guise of security than of the general welfare.

The role of the federal government in science was brought into special focus in 1836 by the then enormous bequest of $500,000 by English chemist James Smithson to the United States of America to establish a Smithsonian Institution "for the increase and diffusion of knowledge among men." Eight years of debate followed in which the idea of a national university was considered and ruled out. Instead, the federal government, in 1844, created an independent institution, under a board of regents, thus avoiding the political problems of whether the government should support a museum, a library, or scientific projects. The institution subsequently became a major force in promoting science, e.g., astronomy, in the U.S.

To review the historical development of the federal role would be interesting, but for present purposes it is sufficient to note that in case after case the federal government became involved only in response to specific needs which could not be met by private interests or state governments. In addition to the Patent Office, the Coast and Geodetic Survey, and the various explorations which we have mentioned, work was undertaken, for example, in 1818 on weights and measures, leading eventually to the creation in 1901 of the modern Bureau of Standards.

In the same spirit the Army Medical Department was created in 1818, the Naval Observatory founded in 1842, and meteorological work was begun by the Army Signal Corps in 1870. However, only the Smithsonian Institution looked to the development of science per se.

Agriculture—A New Departure

The scientific organizations and activities discussed above dealt with quite specific problems. Enlisting science to serve the farmer is a much more general problem. It involves not only basic research but research directed to a seemingly infinite array of crops, soils, climates, and pests. It involves the education of the farmers, the transfer of research results to the farmers, and the transfer of practical experience to the schools and research stations. Finally it involves the economics of the U.S. farm system as a component of the U.S. economy and a world market for food and fiber. The problem was and is to adapt and even to create a productive system based on a large number of independent producers.

How the U.S. coped with that problem is a great success story which is frequently cited in discussions of industrial policy. First of all, though, we should note that the solution was not designed by scientists or systems analysts, and it was not conceived as a whole. Rather, it was born of political pressures as, first, the farmlands of the west were settled and farmers encountered problems for which they were unprepared and, later, when output could no longer grow by opening new lands, by the need to enhance farm income through higher productivity.

The ingredients of the system were not dissimilar from those faced now in industrial development. The first item of infrastructure is the training of people with suitable skills. This was undertaken in a revolutionary way by the Morrill Act of 1862, an "act donating public lands to the several States and Territories which may provide colleges for the benefit of agriculture and the mechanic arts." Morrill said the act envisioned not agricultural schools but "colleges in which science . . . should be the leading idea." Thus were born the land-grant institutions which, though they took many forms and took decades to develop, were devoted to practical problems and to public service rather than to the training of an educated elite.

At the same time Congress established the Department of Agriculture. In establishing a Department of Agriculture and granting public lands for colleges, Congress proceeded on the assumption that its power "to lay and collect taxes . . . for the common defense and general welfare" obviously warranted federal sponsorship of scientific research. Thus, a new era was opened. The Department of Agriculture began to undertake research of various sorts but not without opposition; in 1881 the journal *Science* suggested that agricultural research be turned over to universities and private organizations.

A different solution evolved—a series of bureaus directed toward practical problems: bureaus of entomology, animal industry, and plant industry. They were the core of the research program and pioneered the idea of a stable corps of scientific personnel which could be shifted to various problems as they arose. As a source of trained personnel they looked to the universities, which also collaborated in research. Since, for the most part, those universities were the "cow colleges" set up under the Morrill Act, agriculture in effect evolved a special university system.

The ties to land-grant colleges meant dealing with states, and still another facet evolved. Because the bureaus became collectors of local information and got closer to local problems, they became central information repositories. At the same time, success in their scientific effort made them central sources of information. Finally, the bureaus inevitably became involved in providing other services to the farm community.

These trends culminated in a second major federal step—the Hatch Act of 1887, which created state experiment stations funded in part by the federal government. Since each was attached to a land-grant college, the ties to the states were tightened, and the Department of Agriculture became the focus of a system of semiautonomous research institutions permanently established in every state.

The last feature of the agricultural innovation apparatus came into place as the department became an institution for popular education, as well as of research and regulation. To produce innovation in the sense of new practices, research results have to get into the hands of farmers, and the farmers have to use them. This is not always easy since farmers are often wedded to traditional ways and are contemptuous of prescriptions emanating from the land-grant colleges. This problem was dealt with by an institutional innovation, the establishment of the Extension Service in 1914, to work directly with the farmers. It carried on demonstrations and brought new varieties of crops and new practices to the attention of farmers. The county agent became available as a consultant and the channel for a wide variety of services to the farmer.

The result of all of this is that American agriculture is now a high-technology enterprise which accounts for a substantial portion of the U.S. export trade. The combination of federal and state governments has led the way in generating a system including basic research at centers such as the Beltsville Research Center and the land-grant universities; applied research at the experiment stations directed toward local problems; an information dissemination system; close links to a university system which is a source of supply for the personnel of the Department of Agriculture, for the various state agencies, and for farm managers; and an extension service which reaches to individual farmers. In addition, the research and education system has developed links to the political system and to the large agro-business community. The political system has evolved incentives of various sorts as well as restraints through regulation, all designed in principle to enhance productivity.

The system is probably the most successful government effort to date in stimulating the innovative process. With farming divided into so many small individual units, it is difficult to visualize such an establishment growing to comparable stature and effectiveness in private hands. The question we must ask is whether there are features of the agricultural experience which can contribute to other needs of society. Other sectors which are similarly diffuse and lack private organizations to lead them include the health care delivery system and small businesses. The idea of an organization analogous to the extension service to serve small businesses which lack research resources and management skills is frequently proposed.

Some Alternatives

The early experience, and particularly that of agriculture, points the way to a variety of potential governmental roles in stimulating scientific and technological creativity and innovation.

1. Government can foster the training of people who can serve to staff the scientific and technological enterprise, both in and outside of the government.
2. By supporting basic research, it can foster the development of knowledge and understanding on which all innovation ultimately depends.
3. It can expedite the transfer and availability of information.
4. It can foster the development of infrastructures, of institutions to do what is needed by the whole society and which cannot be or are not undertaken by the private sector, such as the Coast and Geodetic Survey, the Geological Survey, the National Oceans and Atmospheres Program, the National Bureau of Standards, and so on.
5. It can fund and, in some cases, carry on research, development, and even production of items needed by the government itself, such as military and space equipment.
6. Through patent and copyright laws, tax laws, antitrust laws, regulations of many sorts, purchasing policy, and other indirect means, it can provide incentives or restraints which serve to guide the course of innovative activity in the nongovernmental sphere.

Except in the military, space, and nuclear power spheres, we have very few, if any, examples of the development of end products and their introduction into use by government. Whether there are opportunities for such development now is a matter of debate. Even the question of the extent to which "spin-off" (the transfer of technologies from military and space activities to civilian pursuits) stimulates industrial innovation is unresolved. Since military research and development (R&D) is the biggest single category of federal expenditure, we shall consider this question, as well as the impact of the space program, later at greater length.

A persistent question is whether there are categories of industrial activity which for any of several reasons, such as the degree of fragmentation into small enterprises, or the magnitude of the investment required, need governmental leadership or assistance.

For example, would a "Building Research Institute" be fruitful? Related to this question is whether government can make private research more effective, as is alleged to happen in Japan, through the encouragement of "precompetitive" cooperative research by industry.

As we examine some of these possibilities, it is important to remember that the political inhibitions, which led the federal government to spend a total of only $40 million on R&D in 1939, and none at all on research in universities (other than agriculture), vanished after World War II. By 1981 the federal government spent $33.8 billion on R&D (compared with $35.9 by industry), of which $6.8 billion was spent in colleges and universities and $15.9 billion in industry. Since only 33 percent of the federal expenditures were civilian related and 66 percent were defense or space related, the latter deserve special attention in weighing the governmental impact on innovation.

ACADEMIC SCIENCE AND GRADUATE EDUCATION

Before 1940 there was little interaction between the government and universities other than in agriculture. However, it became apparent during World War II that academic scientists provided a major impetus in developing radical new technologies such as radar, the proximity fuse, the nuclear bomb, and new materials. In doing so they called on scientific knowledge which had only recently emerged from basic research laboratories. As a consequence, when the war was over, the armed forces, notably the Office of Naval Research (ONR), funded research in universities and, with it, the training of doctoral students in order to build a foundation for the future security of the country. The ONR program supported a very broad spectrum of work in the physical sciences and mathematics. At about the same time, the National Institutes of Health (NIH) began their extramural program which presently supports almost all biomedical research in universities. The Atomic Energy Commission (AEC) supported nuclear physics, nuclear engineering, and high-energy physics research in universities.

Starting from that base, basic research in all fields has become an accepted responsibility of the federal government. The National Science Foundation (NSF) was set up for this explicit purpose in 1951. However, a number of agencies continue to share in the support of basic research, each cultivating areas believed to be related to the long-term progress of the areas germane to their mission. The 1984 budget, for example, planned expenditures of $6.3 billion on basic research. The distribution of this expenditure among agencies is shown in Table 1.

TABLE 1. APPROXIMATE EXPENDITURES FOR BASIC RESEARCH, 1984

Health and Human Services	$2,200
National Science Foundation	1,100
Department of Energy	1,000
Department of Defense	775
National Aeronautics and Space Administration	660
Department of Agriculture	380
Other	250

Basic research augments the knowledge pool from which scientific and technological innovation draw. Since the results of basic research are disseminated as widely as possible, the benefits of this knowledge are not retained by any single enterprise or industry. They cannot even be retained nationally, but constitute a world treasure which is basic to progress everywhere. It follows, though, that basic research provides benefits only to those who are prepared to make use of it. In itself basic scientific advance does not produce scientific innovation as defined at the beginning of this chapter; however, if the other elements of the innovative process are coupled to it and are sufficiently skilled, it is an essential part of the innovative process, especially for the most radical departures. We have had dramatic examples in the utilization of semiconductor physics in hand calculators and especially in the silicon chips at the heart of computers.

The chief issues involving academic research revolve around the choice of fields, institutions, and projects to be supported. On the one hand, one might select projects as much as possible by the scientific merit of a proposal as judged by peer review. On the other hand, one might temper those judgments by the desire to build or maintain institutions or to achieve other goals. For example, since research experience is at the heart of the education of doctoral scientists, concentrating the research in a small number of elite institutions would expose only a small number of students to the best research minds and techniques. Therefore, it is argued that by carrying on high quality research in a large number of institutions ("spreading it out") the educational benefits will extend to many more students.

A related question concerns the choices among fields. From a purely academic point of view, one looks for areas ripe for scientific advance. But shouldn't other fields be supported because of their continuing importance and relevance to health problems, to environmental problems, or to industrial progress? If active research programs are not maintained might they lose their vigor? They may be needed because their practitioners are essential to industry (for example, metallurgists); other fields may be needed as a base in applied science in defense or industry (for example, materials science). These debates will continue, and one of the great virtues of our pluralistic system is that the various government agencies approach such problems from differing points of view, the results of which can be compared for effectiveness.

In any case, by all measures, such as Nobel prizes awarded, publications cited, or major discoveries made, basic research in America is thriving. There is certainly room for improvement, but if there is a lag in scientific and technological innovation, it is not likely that it results from a deficiency in basic science.

The picture is not quite so clear with regard to either the condition of graduate education or the governmental role in producing doctoral scientists. In one way or another most natural scientists are supported in graduate school by such devices as research assistantships, teaching assistantships, traineeships, or fellowships. The availability of teaching assistantships is chiefly related to the number of undergraduates who take courses in any given field; as a consequence the distribution is not related particularly to the demand for trained manpower. The situation with respect to research assistantships is somewhat better. Their number is roughly proportional to some mixture of faculty research interest in a field and the interest of sponsoring agencies. While that results, in some cases, in training people in fields of great academic but little commercial interest, such as high-energy physics, and possibly directing people out of fields of great applied importance but lacking the highest intellectual stimulation, for the most part, this mode of support has served the country well.

In particular, people have been trained to staff rapidly growing fields such as aerospace engineering or solid-state electronics in the 1960s and computer sciences in the 1970s and 1980s. There are those, however, who feel that in the process talented people were diverted from such areas as machine design and production engineering.

Traineeships and Fellowships—The government has been very ambiguous about the direct support of students or programs to improve the quality of scientific education or the strength of graduate departments. The Atomic Energy Commission supported a program of fellowships to train students in areas important to the development of nuclear energy for a number of years in the 1940s and 1950s. Beginning in the 1950s, NSF offered around a thousand fellowships each year based on the merit of the candidates and their geographical distribution, but for a variety of reasons that program was largely abandoned by 1983. The largest direct fellowship program has been that of NIH. It has been a key ingredient in building up the very large and very successful program in biomedical research in the United States.

Direct support of students has also been undertaken by the government through traineeships. These are coupled to institution building and the maintenance of stable research organizations in universities. The traineeship grant is therefore coupled to a research program. The recipients are chosen by the institution on a merit basis to study and do research in designated fields. As of 1983, NSF had given up its traineeship program, but that of NIH was thriving.

Both fellowship and traineeship programs were based on the idea that it is in the national interest to encourage the most able students to study in areas important to health, national security, and the economy. The approach, which subsidized students going into science and engineering, has been superseded by the argument that the salaries in these fields are good enough to make the expense of advanced study a good personal investment. Unfortunately, the reward occurs so much later that students who have few resources may not be able to continue.

Centers of Excellence—Lastly, the government has undertaken sporadically to assist in building academic departments and new centers of excellence. Such programs have been undertaken by the

National Aeronautics and Space Administration (NASA), Department of Defense (DOD), Atomic Energy Commission, National Institutes of Health, and the National Science Foundation. Most programs have been targeted on areas of science and engineering of interest to the agency. A variety of such programs continue, e.g., the center grants of the National Institute for Environmental Health Sciences (NIEHS). NSF and DOD have set out to broaden the base of American science by strengthening second-rank institutions and departments which show promise of rising to first rank. However, as of 1983, the NSF program had been abandoned; efforts by other agencies were undertaken only as a secondary part of the research program.

In sum, the funding of basic research, and especially that carried on in universities, has become a government responsibility. At the present time it is, for the most part, in a healthy condition, and it is hard to relate any deficiency in industrial innovation to a weakness in this sector.

Whether the supply of a highly trained work force is adequate in the critical fields is less certain, but it is safe to state that work force shortages have not been especially evident. More subtle questions, such as the quality and nature of their training, and the balance among doctoral candidates, baccalaureate students, and students in technical and vocational schools, may be important, but there is little evidence one way or another. In any case, this is not a matter the federal government has addressed directly.

DEFENSE AND SPACE

During the two decades from 1964 to 1984, defense accounted for over half of the federal R&D expenditures in every year. In 1983 these expenditures, 66 percent of which are devoted to the development of military hardware, amounted to more than $24 billion. Another $6.5 billion was devoted to space, principally for the development of rockets and space vehicles. Since most of this R&D and all of the resulting production are carried on in industry and since both defense and space strain the existing limits of technology in their requirements, the degree to which they stimulate civilian industry is an important question. Perhaps more important is whether the degree of stimulation can be notably increased.

During the 1960s it was widely believed in Europe that the experience of American industry in developing and producing sophisticated electronic and communications equipment, new materials, advanced construction and quality control methods, etc., was providing the U.S. with a widening qualitative technological superiority. This outlook gave rise to the perception of a "technological gap" in the 1960s and the fear that this would ultimately result in American dominance. These fears, dramatized by Jean Jacques Servan-Schreiber's *Le Defi Americain,* were so vivid that "the gap" became a political issue which colored relations between the U.S. and Europe for some years.

With the reemergence of intense international competition in high technology, the "gap" has receded as a political force, but the question of the impact of defense and space on industrial innovation is still important. One can look to many possible sources of stimulation by the military and space programs.

1. Both DOD and NASA have supported exploratory research and applied research in both industrial and university laboratories. DOD is our most experienced supporter of research and, in general, has been a very effective sponsor. Its program managers generally could relate their efforts to long-term needs. The NASA program was similar but not as broad. Their long-term interest in high technology in such areas as electronics, aeronautics and astronautics, ships, communications, materials, fuels, etc., provided them with the insights needed to judge the quality and appropriateness of applied research activities. They had the ability to respond quickly, and they understood the value of groups of scientists working together on related problems. The DOD research directors had a degree of venturesomeness which was extremely valuable to the health and progress of U.S. science and technology. The Advanced Research Projects Agency (ARPA), for example, inaugurated a very successful program in materials science.

2. DOD has recognized the desirability of contractor owned technology in its policy and has taken the enlightened step of recognizing R&D on future products as a legitimate cost of doing business. It has done so by allowing independent research and development (IR&D) as a part of overhead. It is a way of hitching the company's commercial interests to government programs.

3. In particular, DOD has supported generic technology pro-

grams to improve the technological base for the production of military equipment by having a supply of advanced technologies "on the shelf." These have included areas such as welding, automated assembly, techniques for forming and cutting metals, and many others. The efficacy of such programs has been a matter of debate, but they clearly contributed to the electronics, computer, and aerospace industries. They deserve further attention since generic technology programs are frequently proposed for civil industry.

4. DOD and NASA have both supported quality and reliability improvements through the development of advanced, nondestructive inspection techniques such as X-rays and ultrasound.

5. Through their large R&D programs DOD and NASA have expanded in high-technology areas the pool of scientists and engineers, many of whom migrate into civilian industry beyond the numbers which would have been available in their absence. On the other hand, they have driven up the cost of research and have drawn people out of less glamorous, less well financed fields financed by industry alone.

6. Above all, by providing a very large market for the most sophisticated, high-technology products, manufacturing skills have been developed. Defense and space programs, for example, have provided the first market for the largest, most capable computers at each stage of development of the industry. They have provided a large-scale demand for miniaturized devices with very high reliability.

In a general way in the areas of electronic devices, communications, computers, light-weight structures, etc., defense and space force the state of the art in many fields through their requirements for high-quality, high-performance production items and their willingness to pay for them. Needless to say, cost plays only a small part relative to performance in military and space equipment so that it has frequently been noted that many of the best production sources for defense and space have not successfully entered the commercial market. That is true, of course, for some, but others, such as General Electric and Westinghouse, also have large civilian businesses, and several industries derive their commercial positions from prior military research, development, and production.

THE AIRCRAFT INDUSTRY

It is probably reasonable to assert that the predominant position of the U.S. aircraft industry in the world market would not have been achieved without direct support by the government. Systematic aeronautical research in the U.S. began with the creation of the National Advisory Committee for Aeronautics (NACA) in 1915 and the laboratory set up at Langley Field. In the succeeding years, its wind tunnels became the source of progressively more efficient wing sections, and it led the way in research for the aircraft industry in aeronautic engineering, propulsion, and structures. Eventually NACA evolved into NASA, but its leadership in aeronautical science continued. Not only did the government carry on research in its own facilities, it funded most of the university work in aeronautical science and engineering and most of the R&D carried on by the aerospace industry.

Above all, it purchased thousands of aircraft, ranging from high-performance fighters to long-range, heavy-lift cargo planes. It bought specialized planes such as the U-2 and the RX-70 (a very high-altitude, very high-speed reconnaissance aircraft). It bought helicopters of all sizes and capabilities. Jet engines came to maturity in military aircraft before they were used in civilian aircraft. Substantially all commercial engines are derived from military engines, and, until recently at least, most commercial aircraft were derived from earlier military designs. Most of the new alloys and structural features were first employed in military planes.

In short, military aircraft was the large-scale proving ground for most of the innovations in the industry, and it seems apparent that innovation in the commercial aircraft industry is, in large part, built on prior innovation in aircraft for military purposes.

COMMUNICATIONS SATELLITES

What has just been said about the aircraft industry is also true of perhaps the most important communications advance of modern times—the communications satellite. The rockets which carry them aloft were developed under the aegis of NASA and DOD, and the early versions of the satellites themselves were developed and built for NASA and DOD. DOD is still the largest purchaser and user of communications satellites. It is plain that except for DOD and NASA the industry would not exist, and its continued progress still relies heavily on the government programs. One additional government contribution should be noted since it may also be important in other connections. As a major purchaser of satellite communication channels the government helps to maintain the economic viability which makes commercial operation feasible.

Spin-Off—In sum, the Department of Defense and the National Aeronautics and Space Administration have played a crucial role in the innovations essential to the aircraft and communications satellite industries. They have also made significant contributions to the innovation process in the electronics industry. In all of these cases applied research and technology development related to end products which were directly relevant to their own mission. Despite the success of the jeep in civilian use, there is little evidence for spin-off of the technology to commercial enterprises in general, although NASA tried very hard to effect such transfer of technology. It is hard to know how much the early DOD and NASA experience with integrated circuits or miniaturized equipment eventually affected the consumer electronics market.

Nuclear Power

The only example known to the author of a deliberate effort by the government to establish a new industry is the attempt by the AEC to bring a nuclear power industry into being. The Oak Ridge National Laboratory not only carried out basic and applied research, but it studied the properties of materials for power reactors and designed and built successively larger nuclear reactors of a variety of designs. At the same time, AEC instituted collaborative programs with industry, which was also involved in building nuclear propulsion units for submarines. In these ways the development of industrial skills was heavily subsidized. In 1957 the Shippingport demonstration reactor was built at government expense, industry being charged only for the value of the steam produced. After a number of further subsidized demonstration plants, nuclear power became fully commercial in 1968.

However, even since then R&D in materials and reactor design have continued in government laboratories and received government subsidies in industrial laboratories.

The result of this governmental effort has been to put together a nuclear power industry which leads the world. Both France and the United Kingdom have adopted American pressurized water designs. On the other hand, with much less governmental investment, they and West Germany have become competitive with the U.S. in the world market. Whether the billions of dollars invested in establishing a nuclear power industry will eventually be regarded as a wise expenditure remains to be seen.

Direct Support of Industrial Ineffective Development

For the most part, the government has not been involved in the selection and management of technological development programs aimed at commercial applications. When it has undertaken demonstration projects to stimulate new industries, the results, with the exception of nuclear power, have been disastrous. For example, the Morgantown, West Virginia, personal rapid transit project assumed that technical demonstration was all that was required. Such demonstrations tend to ignore such necessities as capital, production, distribution, servicing, and repair. Success requires an experienced enterprise with a high stake, and these considerations received little attention in planning and executing the Morgantown project. It eventually sank without a bubble, carrying with it well-intentioned millions of dollars.

Another such government effort was the attempt to develop a commercial supersonic transport aircraft in the 1960s under the aegis of the Department of Transportation. Fortunately, the program was abandoned before we had been committed to a production program, but not until after the expenditure of several hundred million dollars. The British and French governments were not so fortunate, and the Concorde has continued to be a costly white elephant without even producing national prestige.

The vulnerability to arbitrary oil pricing and even to actual interruption of the oil supply, demonstrated in 1973 and again in 1978, led to direct action to develop alternative energy supplies. In this case the government funded research on alternatives such as solar energy and the construction of pilot-scale solar energy plants. It also offered tax credits for the installation of solar heating equipment. In addition, it fostered applied research on the gasification and liquefication of coal and, through the Synthetic Fuels Corporation, collaborated with industry in the construction of large-scale plants to produce liquid fuel from coal and oil shale. The subsequent drop in the price of oil caused the abandonment of most of this program, and it remains to be seen whether it will not be needed in the long run.

Other governments have gone further than we have. Putting aside the matter of nationalized industries in France and England, France has attempted to accelerate the development of high-technology industries through such means as the Plan Calcul, a scheme to promote the manufacture and use of advanced computing equipment. England has founded the National Research and Development Corporation whose greatest success has been the antibiotic cephalosporin; its commercialization of the Hovercraft has been marginal at best. None of these examples has been sufficiently successful to warrant emulation.

Japan, of course, is most often cited. However, the Japanese government has not entered the commercial market directly. Rather it has attempted to encourage and assist industry through various collaborative schemes. It has also supported R&D in areas targeted for commercial development, such as automated manufacture and robotics and computers. It remains to be seen whether such targeted efforts will be successful in the long run.

Health and Biomedical Research

One area whose success derives very largely from government support is that of the biomedical sciences. Beginning in the 1950s, the fields of biochemistry, molecular biology, biochemical genetics, immunology, virology, and so on were almost entirely supported through the National Institutes of Health. This so-called genetic revolution, which transformed our understanding of health and disease, grew exponentially, and the federal government financed the training of people, the conduct of research, and the providing of jobs for the people.

For many years this basic research had little impact on health statistics, the treatment of disease, or the pharmaceutical industry. Now all of that has changed. The mode of action of pharmaceuticals is being understood, and they are acquiring a rational foundation. Organ transplants are made possible by the advances in immunology, and, above all, a new industrial frontier in genetic engineering has been opened which will have great impact on both health and agriculture and, in the long range, other segments of industry.

The government has not participated in the commercialization of the products of the basic, applied, and clinical research it fostered, but it can legitimately claim to have set the stage for the most dramatic and radical new scientific innovation of our time.

Indirect Government Roles: Incentives and Restraints

Many analysts are of the opinion that while government action in supporting research and advanced training, as well as providing backup services, is essential to the innovation process, the greatest impact of government comes through its general social and economic program. Most often mentioned for their impact on industrial innovation are:

1. inflation and interest rates,
2. tax policies,
3. environmental health and safety regulation,
4. the patent system,
5. the disposition of patents resulting from federally funded research,
6. antitrust policy,
7. federal procurement policy,
8. policy toward small and innovative firms, and
9. transfer of technology from federal laboratories.

Inflation and high interest rates are commonly thought to inhibit innovation and reduce productivity growth by reducing the rate of return on new ventures. Though inflation obviously does reduce the rate of return, Harvard economist Dale W. Jorgenson notes that the real rate of return (above inflation) has been comparable in the period around 1980 to the average of the post–World War II period so that it, in itself, is not the major problem. Nonetheless, if capital flow into innovative ventures is to be maintained, government policy should insure that the real rate of return remains adequate despite inflation.

Nearly all analysts focus on the lag in capital formation in the United States, and Jorgenson traces the decline in productivity growth in recent years largely to this cause. The essence of this argument is that the lag is not so much in generating technology as in the fact that we do not use the R&D which is already available to us. In this case the highest priority for government action is to take such steps as will stimulate capital formation and, especially, the generation of venture capital. The principal governmental tool for doing so is through tax policy. (A detailed discussion of tax policy is beyond the scope of this chapter.) However, among the measures most commonly suggested to spur capital formation are accelerated depreciation schedules for investments in plant and equipment and the use of replacement costs rather than historical costs. A variety of other tax moves to stimulate investment, among them reduced capital gains taxes, have been advanced.

One of the interesting features of scientific and technological innovation is that a very high proportion (perhaps 50 percent) originates with individual inventors and small enterprises. Therefore, a particular goal for government policy should be to stimulate both the formation and the health of such enterprises. Direct investment in new ventures, which involves detailed commercial choices, is not well suited to government action. But tax and security laws can be rewritten to encourage equity investments in small companies, and the government generated red tape involved in running a small new business can be cut back.

Accelerated depreciation schedules for equipment and structures would provide incentives for private R&D as well as for investments in facilities and equipment for production. Other incentives to private R&D might include tax credits for R&D expenditures, and these stimulants have been recommended by some. However, it is not evident that this is a serious problem since R&D expenditures by industry have been rising steadily. For example, from 1973 to 1983 they have risen from $13.3 billion to $44.3 billion.

Although there are instances where regulation has been a stimulant, it is, by and large, seen as a restraint on innovation. For example, it is alleged that capital diversion to meet the requirements of environmental health and safety regulations constitutes a serious brake on innovative capacity. It is also felt that the uncertainties engendered by regulation and possible changes in regulations inhibit the establishment of new ventures. There is no doubt of the nuisance effect of regulation, but the overall effect is less certain. Jorgenson points out that the capital diversion is only a very small part of the total rate of capital formation. In another study, economists Gregory Christiansen, Frank Gollop, and Robert Havemann concluded that the impact of environmental health and safety regulations on macroeconomic performance and productivity growth was not important. Nonetheless, even if the average effect is not large, the impact on individual sectors and particular enterprises may be very important. In the future the impact on innovation needs to be considered in developing regulations, and this needs to be done separately for the various sectors.

Still another facet of governmental policy arises from antitrust laws. As presently interpreted, collaboration between competing companies is very difficult at any level. The question which arises is whether such restraint is necessary to sustain competition, especially when the domestic industry is faced with vigorous foreign competition. To what extent should members of an industry be allowed to cooperate in R&D which will enlarge the technological capacity of the entire industry? It is widely believed that the current barriers to collaboration between competitors at this level are excessive.

Patent protection for inventors and entrepreneurs was written into the Constitution and has always been the principal guarantee to inventors and users of inventions that a return could be realized. Yet most patents are actually weak, and the time and expense of establishing their validity through litigation are excessive; in addition, the uncertainties involved increase the risk to the entrepreneur. Most observers agree that the innovative process could be improved through patent reform which would strengthen patents, once issued. This, however, would require greater vigor in the examination of patent applications.

Related to this is the question of the disposition of patents issued to contractors or grantees resulting from government funded research, as well as those patents on inventions in government laboratories. This has been a subject of debate since at least 1945, and the practices vary widely between agencies. DOD grants title to its contractors for patents taken out as a consequence of federally funded research, recognizing that the cost of reducing inventions to practice, setting up production lines, and marketing the product is usually much greater than the original research. Moreover, it usually involves substantial risk.

The other point of view, that the results of publicly funded research should remain in the public domain and be made freely available to the public, characterized the Atomic Energy Act of 1946; the NASA legislation; the Department of Agriculture; the Non-nuclear Energy R&D Act of 1974; and, historically, the Department of Health, Education, and Welfare (now Health and Human Services). Under this policy, the government retains title; only nonexclusive licenses are granted, except under special circumstances. Although a few government patents, such as the aerosol can and frozen orange juice, were successfully commercialized under this policy, the main result was that literally thousands of government patents have never been exploited, thus wasting the possible benefits of the investment in the research on which the patents were based.

Some progress was made in this respect when, in 1963, President Kennedy issued a statement on patent policy which attempted to state a rationale for the diverse patent policies then in existence. It called for a flexible patent policy rather than a uniform one, balancing the various objectives: to stimulate R&D, attract contractors, avoid monopolization, and recognize the equities of both the government and the contractor. Under a number of circumstances the government would take only a license to inventions, leaving ownership and commercial rights to the contractor, who was thought most likely to develop the inventions for commercial use and practical benefit to the public. In 1971, President Nixon reaffirmed the Kennedy statement but amplified it to encourage agencies to grant exclusive licenses to government owned patents where necessary to stimulate commercial applications of these patented inventions. In addition, agencies working in areas of public safety, health, or welfare, which were normally instructed to seek title, were encouraged to consider leaving title to contractors in exceptional circumstances.

In 1980, this trend culminated in Public Law 96–517, which provides that universities, nonprofit organizations, and small businesses could elect to retain title to inventions resulting from government funded research, subject to certain disclosure requirements. It provides "march-in" rights for the government if an effective effort to achieve practical application of the invention is not made in a reasonable time. This law also defines the conditions under which government owned patents, both those resulting from grants or contracts and those emerging from government laboratories, may be licensed. It allows for exclusive licensing when it is necessary to call forth the investment of risk capital and expenditures to bring the invention to practical application.

The most recent step has been a memorandum from President Reagan encouraging all agencies to transfer title to the R&D contractor when it is consistent with their enabling legislation. Under this policy, the Department of Defense continues to transfer title routinely to large industries while the Department of Energy grants only licenses.

Today the federal government is industry's largest single consumer and customer. Consequently, it has frequently been suggested that government procurement policy can be a potent tool in stimulating innovation. This has already been the case in the aircraft industry, the satellite communications industry, the computer industry, and many elements of the electronics industry. Federal procurement played an important role in development of both transistorized IBM computers and Xerox copiers. The question is whether or not it can also be effective in the housing industry, the automotive industry, or other consumer industries. DOD has attempted such stimulation by replacing the usual construction specifications with performance specifications for military housing. However, it has had no appreciable impact, probably because the volume of purchases was too small and specialized. Despite the success of the jeep in the civilian automotive market, the same lack of impact would probably apply to any attempt to influence that market via military purchases. The outlook for a widespread government role by this route does not seem promising, but where incentives for the development of advanced products can be supplied, the attempt should be made.

GOVERNMENT LABORATORIES

Aside from its role in funding and guiding research and development in universities and industries, the federal government operates several hundred laboratories. These include the NIH;

the large NASA centers; DOD laboratories, ranging from important scientific centers like Naval Research Laboratory to a large assortment which have no mission; agricultural laboratories; the Bureau of Standards; and the national laboratories of the Department of Energy (actually managed under contract). Together they represent an important national resource. Where they have had a role in supporting major missions of their parent agency (for example, advancing space travel), they have been successful. But frequently their absence would not be missed. In any case, they have not contributed significantly to industrial innovation in general.

Some have made the attempt. Former AEC laboratories, notably the Oak Ridge National Laboratory (ORNL), have reacted to the decreased emphasis on nuclear reactor development by attempting to broaden their base and become a general purpose national resource. ORNL, for example, has built a very strong biology division based on its expertise in the medical and biological uses of isotopes. When President Johnson sought to push the desalination of water to meet the needs of arid regions, ORNL became the principal R&D center for the effort, but nothing of real industrial importance emerged. Nonetheless, these efforts have resulted in broadening the research base and made it possible for some of the laboratories to recruit and hold high-caliber scientific staffs. To date they have not contributed significantly to industrial innovation.

Conclusions

Any discussion of scientific and technological innovation in the United States must take account of the major role played by the government in that process, both in stimulating and restraining it. The government plays a predominant role in both the conduct of basic research on which the future of technologically advanced industry depends, and in the education of technical people at all levels. In areas related to its responsibilities, notably health, agriculture, energy, defense, and space, government is also the principal supporter of applied research. Through the patent system it offers essential protection to inventors and entrepreneurs. It carries on research in areas upon which industry depends but which would not be sponsored by the private sector.

Lastly, the government organizes information services and encourages the transfer of knowledge and technology.

It has not successfully stimulated the development of particular industries except as a by-product of large-scale defense, space, and nuclear programs. However, in at least one area, agriculture, it has been the nerve center around which a highly successful, scientifically advanced, and very productive agro-industrial enterprise was built.

In the context of scientific and technological innovation to develop industry and promote the general welfare, the present question is whether it can or should attempt to promote and stimulate designated industries and play a role such as that of the Ministry of International Trade and Industry in Japan. One must conclude that we have not yet found a way to do so.

On the other hand, the agricultural experience suggests that it might be possible to provide scientific, technological, and managerial assistance to small industries whose creation and survival are essential to innovation.

Another question is whether government support should be given to programs of technological development as opposed to scientific research. The goal of such programs would be to have a wider variety of technologies "on the shelf" for adaptation to particular tasks by civilian industry. However, it has not yet been established that such development is effective in the absence of specific challenges and goals.

Aside from specific interventions, the government has a tremendous indirect role. It can stimulate and guide capital formation through its tax and economic policies. It can shape the direction of innovations through its regulatory policies. It affects the conduct of innovation through antitrust policy and the steps it takes to promote small, innovative ventures.

In sum, the government cannot step aside. It has long since been too much involved. It should still lead the way, seeking out the things it must and can do, and providing incentives for action by the private sector in the areas which industry does more efficiently.

Robert A. Frosch

3

Improving American Innovation:

The Role of Industry in Innovation

A thing of beauty is a joy forever:
Its loveliness increases; . . .
Keats, *Endymion*

Introduction

It seems clear that it is the aspiration of the United States to continue to produce a flow of products, new products, which will dominate world markets and make profit and reputation for the manufacturer. Our traditional world business role has been to produce new products with a reputation for being the most innovative and advanced in their fields; generally for a market-dominating combination of quality and price.

All of the adjectives in the previous statement seem to be important in maintaining the competitive position. Reputation for past products is not independent of the ability to acquire sales with new products.

ROBERT A. FROSCH *is vice president of the General Motors Corporation, in charge of Research Laboratories. Previously, he was administrator of the National Aeronautics and Space Administration and president of the American Association of Engineering Societies. Dr. Frosch is on the advisory committees of several national research institutions and universities and serves as a trustee of the Woods Hole Oceanographic Institution and Engineering Information, Inc. He has published numerous articles in scholarly journals and for research organizations.*

Apparently our nearly effortless ability to provide this continuing flow of innovative products and to maintain our reputation for them and, thus, our sales seems to have declined, perhaps almost to have vanished in some key areas. We perceive our reputation and competitive edge to have slipped badly and to continue to slip in world markets. Worse, foreign competition has invaded a number of U.S. market sectors (e.g., automobiles, consumer electronics, and a variety of household appliance lines and industrial goods) with products that are perceived to be, and frequently are, technologically more advanced and of higher quality than our domestic production. Even in fields in which Americans are the original inventors and developers, we are overtaken within a short time by foreign versions of our original ideas. We no longer necessarily dominate markets, even with our own inventions.

This difficulty persists, and may be growing, despite an extremely strong scientific enterprise, and there is evidence that our research in advanced engineering continues to be very strong. Something appears to have changed, and it is not completely obvious what it is.

What is the nature of innovation, its environment, its enemies? All these concerns should be considered in determining what innovation should be and how it should work. This will lead, implicitly at least, to thoughts about what could be done to improve our position.

The problem will be divided into several parts: the research and development (R&D) process (where the ideas come from), the innovative process (how things get to market), the organization (its alternatives and its effects on the innovative process), and, finally, some discussion of external factors and sectoral relationships (who does what, and with which, and to whom).

The R&D Process

> *To be, or not to be: that is the question:*
> Shakespeare, *Hamlet*

The research and development process is the generation of ideas and their development into technologies that can become major contributors to the innovation of new products. It is fundamentally a creative process in which people, who frequently have little or no product development motivation, try to build a body of knowledge in subjects of interest to them and try to turn that knowledge into new technological capabilities building upon previous technology.

Research and development must be somewhat distinguished in their motivation. Research people are concerned particularly with ideas and knowledge and not necessarily motivated to produce means for the use of these ideas other than as part of providing the technology required for their own search for knowledge.

Development people, on the other hand, are frequently motivated by an interest in making things work, in turning knowledge into some useful technique or capability. We distinguish between two kinds of development: technology development and product development. Technology development usually arises only from a motivation to make things work and may not be well coupled to the creation of a salable product. Indeed, as will be seen, deciding when to shift from making something that works in principle to making something that works as product may be an important timing choice in the innovative process.

All aspects of the research and technology development process are creative processes in the artistic sense. They are difficult or impossible to schedule because they depend upon a flow of ideas and inspiration much more than on systematic processes and the carrying out of procedures. Procedure and process are important in demonstrating the validity and usefulness of ideas, but in the early R&D stage the really good ideas are much harder to come by than the processes for testing them.

Because research is an artistic process, having a research laboratory is rather like owning a stable of poets. People engaged in this work are likely to be individualistic, somewhat "unreliable" in their rate of production (even in their behavior), and rather variable in their performance. They vary greatly in their characteristics and, on the whole, are likely to impress systematic managers as being extremely untidy and difficult to deal with.

Research scientists work immersed in a sea of ideas generated by themselves and those around them, as well as perceived by them in the scientific and other literature. Relatively little appears to be known about the sources of perceived problems and ideas. It is an extremely individual creative process.

An aspect which is frequently disturbing to systematic management is the problem of irrelevance. It is easy to define those possible topics of research that are clearly relevant to any particular subject. It is much more difficult to decide what areas, apparently irrelevant to the subject at hand, will later turn out to be extremely important. For example, in the 1950s it would have been clear to many people that improved glass was important to the future of better optical devices, but it was not apparent to anyone that consideration of the statistical distribution of energy levels and a phenomenon originally called "negative temperature" was of great interest for the improvement of optics. Nevertheless, the latter set of preoccupations led to the laser.

The importance of the seemingly irrelevant is one of the most difficult aspects of the management of industrial research. The most important innovations, insofar as they depend upon new knowledge, frequently depend upon knowledge that arrives from a direction not previously perceived as having much to do with the problem. The best we seem to know how to do is to choose general subject areas that have something to do with the products that may eventually be of interest, to use excellent research people, to give them reasonably free rein, and to expose them to a multitude of ideas about the kinds of eventual products and subjects of interest to the innovator.

This type of discussion of the nature of the research process is not very helpful to managers of industrial research who do not have research backgrounds, but it makes clear why a number of dilemmas surround the problem of how much internal research a corporation should do, how much, if any, contract research, and how much it should depend upon totally external sources for the knowledge and research ideas to be used in innovation.

There are, however, some useful things that can be said. It seems from the prescription above that it is nearly impossible, except perhaps for the federal government or for the largest corporations, to maintain sufficient internal research capability to be able to investigate the obvious relevant subjects and a reasonable population or the unobvious ("irrelevant") research subjects to be sure that they are covering enough bets in their early research areas. At the same time, an industrial operation with no research people is unlikely to have sufficient antennae merely to study research produced by others in order to find precisely those ideas that will be important to its next generation of products.

Thus, there is a kind of minimum research "intelligence sensitizing" operation that innovative corporations require. People capable of recognizing what is going on in a subject are unlikely to keep their edge if they are not somehow engaged in the research process themselves. As the late Alfred P. Sloan, Jr., has pointed out, this implies that to have an adequate research operation, there must be within the organization a group of people who are honed and interested in a variety of subjects that may be important to the corporation.

Their work can then be supplemented by a continuing knowledge of the published and openly available work of other industrial research organizations, of universities, and of the worldwide knowledge-producing organizations. Thus a judicious combination of internal work, external communication, and "R&D watching" is important to understanding the nature of the knowledge that may be used for innovation.

The external connections must be more than mere reading of the published literature. They must include some participation in meetings and organized research activities of the communication kind. This is because the delay between new ideas, their informal circulation in scientific communities, and their eventual publication can be sufficiently long so that, if one only reads, he is always significantly late in learning of new, continually emerging research and ideas. This difficulty becomes more acute as time passes since the process of change in science and engineering accelerates, possibly as a consequence of improved technological means for communication. If so, this will continue to accelerate as means of communication become more and more mediated through computers, permitting large networks of people to exchange data and ideas conveniently and inexpensively.

An important way in which an industrial research organization can extend its scope and reach is found in a variety of relationships with other research organizations. Since it is generally not possible for it to have close relationships with competitors, or frequently even with those in related industries, it is logical for the organization to extend its possibilities by relationships with university and government laboratories.

INDUSTRY COLLABORATION WITH GOVERNMENT LABORATORIES

Although there are increasing relationships of collaboration and of use of each other's facilities, which can be developed beneficially for the advantage of both government and industrial groups, there are complications in the case of government laboratories. The principal complication has to do with intellectual property and general government insistence that, when government money is involved, the public right in the intellectual property that results must be captured by the government. This area is changing, but so far this attitude means that industry and government are somewhat at arm's length in relationships where the corporation is interested in keeping all the rights in the intellectual property. There are notable exceptions, especially in those industries where the government is a major customer or where government sponsored industrial research has become extremely important. This is particularly true in aerospace and Department of Defense related industries.

INDUSTRY COLLABORATION WITH UNIVERSITIES

A variety of mechanisms have been developed for useful industry-university relationships in research, and there is a great deal of current activity in the invention of new arrangements. A traditional arrangement is for university faculty engaged in research in subjects of interest to corporations to be retained as consultants by the corporations. While employed as a consultant, the faculty member is an employee of the corporation. This arrangement is used to protect the proprietary rights, if any, of the corporation with regard to things that the consultant might learn from the corporation.

In many cases these relationships go beyond consulting (in the usual sense of advice) into corporate sponsorship of university research. This is normally totally open, with publishable research having no proprietary rights for either party. This relationship with the university can be used to expand greatly the industrial organization's direct access to excellent research upon a wide variety of subjects and topics. In some cases this has resulted in collaborative research where both parties are finally the authors. These arrangements provide a means for the industrial organization to widen and extend its contact with university research and, by judicious sponsorship, to find specialists or professionals with similar interests. It thereby, in effect, extends the actual research on subjects which are either of interest to it or which may fall in that class of potentially important "irrelevant" research.

An additional useful consequence of these relationships is the contact that they provide with students, especially graduate students, with consequent recruiting advantages for the corporation.

These relationships are of a traditional kind and are well understood by most universities and corporations. They do not generally have any but positive effects on both the mission of the universities and the mission of the industrial research laboratories.

Counter Flow from Industry—There has been much less flow of industrial research personnel into university work, except in some instances upon retirement. Not infrequently industrial people teach part-time at local universities or, in rare instances, take "sabbaticals" to work in university teaching and research. This is much less common than consulting and summer work for industry by faculty and students, and it seems to present a number of difficulties from the industrial side, particularly with regard to the continuity of jobs and career patterns in industry.

UNIVERSITY RESEARCH CENTERS AND RESEARCH PARKS

Recently a number of new arrangements have been tried to bring industrial and university research closer together. Many universities have organized research centers around particular subjects they believe to be of industrial interest. These subjects have included large-scale digital circuit integration, robotics and computer aided design and manufacturing (CADCAM), and genetic engineering and related fields of research.

In a number of cases the pattern has been for the university to organize a center and invite a number of industrial sponsors to participate in supporting it. As a return for their sponsorship in money, in kind, and possibly in the direct provision of people and ideas, the industrial sponsors get frequent, early, and somewhat special access to what is going on in the research center. There are no legal proprietary rights in the results, and the research center does not sequester its research for only its sponsors; research results are published in the normal way. However, the intimacy of access, the possibility of general intellectual influence, and perhaps direct influence upon the problems chosen for research (by presenting those of possible interest, which may be different than would otherwise have arisen) seem to constitute a sufficient lure to make the best of such arrangements successful.

In another arrangement, there are several cases, particularly in the digital electronics industry, where a group of industrial firms amounting to a trade association collects funds which are used to sponsor relevant research at a number of universities. Presumably there will be less of the special access relationship in this kind of sponsorship; rather it is a means for a number of industrial firms, which might not individually have the capability to do all the research they would like to do or sponsor, to stimulate that research in the university laboratories. Thus is increased the availability of a pool of ideas and knowledge upon which business may draw for the development and innovation of products.

Another device is the development of an industrial or research park in conjunction with one or more universities. In this situation the university acts as developer of an R&D oriented industrial situation close to itself or as stimulator of that development by others. The desired result is the establishment of a community of innovative firms with which the university may work, both in terms of corporate sponsorship of university research and of interchange of personnel and ideas.

In all of these cases there is an exchange of benefits: the university benefits from financial support and by stimulation for its research from questions relating to innovation posed by industry; the corporation benefits from the stimulation of university research, the availability of knowledge and concepts to which it may have special or intimate access, and, hence, the contribution of research ideas required for product development.

Applied Research and Development

The early stages of development constitute an area that is frequently called "advanced development." In the Department of Defense it is referred to as "exploratory development," with the term "advanced development" being reserved for a slightly later part of the process. In any case, this is the time when a new idea has emerged as a phenomenon, a material, or a process which clearly has relevance to either the corporate product or to the manufacturing processes for that product. The problem is how to turn the idea from a laboratory event or curiosity into some kind of useful engineering or product result—how to develop it technologically.

This aspect of the problem is frequently regarded as "routine engineering," but is far from that. The process of developing a technology from research knowledge is a creative process in itself, because it involves the development of totally new designs and ways for making things based on knowledge not previously in existence or use. Development involves a different kind of process from the basic research process. Because the new knowledge will not be of particular use all by itself, it must be fitted into a matrix of other materials, other mechanisms, and other portions of what will finally be a working technology.

At this stage there is likely to be little constraint regarding product design or cost; as will be seen, that comes later. The concern now is to turn the knowledge into something that can be made to work in a reasonable way, not yet in a way which is refined enough to be product. Because this process involves imbedding the new possibilities in a matrix of older systems and ideas, there is less free play and far more constraint than in the research context, although the process still involves experimentation and inspiration; creative idea generation continues to be extremely important—more important than being systematic.

It is during these early attempts to use new phenomena, materials, or knowledge to develop new technologies and working systems that gaps, errors, and difficulties in the research product (knowledge) are frequently detected. It is here that the things not measured or noticed in earlier research can become important. Frequently there is strong feedback from this early development which leads to refinement and improvement of the original research.

It is important to note that development of this kind is frequently done by somewhat different kinds of persons than those engaged in research. Research people characteristically are interested in carrying through the more constrained process of producing a designated end result from their research results. Development, on the other hand, is frequently more an engineering skill than a scientific one, or even a research engineering pursuit.

There is a particularly interesting process going on in much engineering research today which bears upon the nature of the process of development of research ideas into technology. Many of the traditional engineering technologies, metal casting and sheet metal forming being good examples, can best be described as well-formulated traditional crafts. There is a body of phenomenal, logical, and empirical knowledge, with bits and pieces of theoretical understanding, which forms a well-understood art that can be learned. Skilled engineers in these areas carry a body of knowledge which can be applied not only in product design and development, but also in the integration of new research knowledge into new technological possibilities. However, the process is inherently difficult and limited since the basis of understanding any of these crafts in a broad scientific sense does not exist, and thus the knowledge is not there to extend the processes and techniques into areas that are very different from those in which they have been previously tested and applied. Hence, the process of using new knowledge to expand into new possibilities with these technologies can be very slow and partakes of the character of tinkering, rather than of systematic design and development. Consequently, an important line of engineering research at the present time is the process of putting many of these traditional crafts on a sound basis of scientific and theoretical engineering understanding.

By itself this process does not necessarily provide new technology, but it lays a foundation for rational extension of known technology into new areas of development. Much of this work has been made possible by recent advances in computer technology which allow the application of theoretical and fundamental physical and chemical ideas to areas where sheer mathematical manipulation and computation difficulties would have been a major barrier a few years ago. This is not the development of new scientific knowledge in a fundamental sense, but rather the ability to apply scientific knowledge to rather complex engineering problems in entirely new ways.

This brings us to the point in the innovation process at which a piece of new knowledge has been turned into an actual capability; the idea has been made flesh. Although this is not yet a product, a first step has been taken that may make an innovation possible.

Returning to the laser example, we have arrived at the stage where an understanding of stimulated emission from an inverted-state population has been turned into a laboratory microwave amplifier and then into a radiating light-emitting population of molecules in the laboratory. This is now a demonstration that the understanding can be made to work as a device, but not yet anything that could be used as a marketable product. There are other examples in materials and many other developments where a device can be produced, but it is not yet anything that could be made to be bought or sold.

PRODUCT DEVELOPMENT AND THE INNOVATIVE PROCESS

> *This little piggy went to market,*
> *This little piggy stayed home.*
> Mother Goose

A Digression—At this point I must digress to point out that, for purposes of providing this chapter with a reasonable flow, I have used a traditional model of the innovation process, one which proceeds systematically from research and scientific ideas through their development into working technology and continues their development into salable products, the development of manufacturing technology to make them, the construction of the necessary machines and plants, etc. The reader should keep in mind that this linear time sequence is a grossly simplifying artifice. In the real world this is a kind of technology-push model of the process in which it is the new ideas that lead to the creation of products. It is frequently the case—sometimes it is argued, for no good reason that I can see, that it should always be the case—that it is the product idea which comes first, followed by a development constructed to produce the product; sometimes research is initiated to provide the knowledge that might make the product idea possible. It is clear that ideas for product, ideas for knowledge, and ideas for development can originate in many places and at different times in the process, and all of them stimulate and guide the others.

It is particularly true in the industrial context that knowledge of the aspirations and plans of the firm and what it intends to do in a general, future-product way can be important, if not essential, to the proper guidance and conduct of the internal research and development of the firm. The managers of industrial research in the firm should be knowledgeable of, and sensitive to, the product interests of the firm, but by no means should they allow themselves to be overwhelmed and entirely channeled by those interests, or else they will surely miss the most important "irrelevant" research and even some of the obvious relevant research and development that can lead to really new and interesting products. More will be said about this later when the subject is treated in terms of internal and external technology transfer.

The whole process can be viewed as a sequential line from research to product introduction with a set of nested feedback loops which connect the various parts of the process to each other. This model makes it seem nearly as complex as it really is.

With this in mind, the question of development is approached again, but now explicitly as part of a process intended to lead to something which can be sold as a product in the market. This step is the central portion of the innovative process—the conversion of a working development into a real product. This may conveniently be referred to as product development.

A product is different in many ways from an idea that has been shown to work. In the development of a product, major aspects of the product as a system and as part of a system arise. What previously was only a working device must now be imbedded in an environment somewhere out in society. New requirements arise, including requirements for reliable and safe operation, perhaps for the ability of the eventual product to be maintained in its operating environment, and for servicing when and if it fails.

The question of what it will finally cost to produce the product as eventually designed and manufactured is frequently dominant in product development. Indeed, the question of *when* to worry about cost can be a dominating problem in the innovative process. In many cases the worry about cost of eventual production has become so dominating a question that perfectly promising research and advanced development possibilities are destroyed because of too early a set of assumptions about the cost effects of their later development.

It is frequently assumed that if something is difficult and complex in the laboratory, it will also be difficult and complex and costly even after its eventual development. As stated, often this kind of business attention too early in the development process can be a destroyer of useful results. This is true in spite of the fact that there is abundant historical evidence that predictions of cost, and even of market, for brand new devices were not only wrong, but wildly wrong when made early in the process. The past twenty years of history of costs of computer hardware and software need only be suggested to make this point. (The same needs to be said about the nature of the market, and that point will be picked up somewhat later.)

Nevertheless, it is necessary to have some concern about cost in a number of stages in the process. For example, if a new material being developed in the research laboratory depends in some essential way (and one had better be sure that it is essential and not merely a curiosity of the current stage of the research) on some other material of intrinsically limited quantity, a rare element, for example, then it is worth discussing cost before going too far with the research. However, it may be cautioned that early estimates of required quantities have a bad habit of being incorrect. What is more difficult to remember or predict is that learning how to do something and understanding what one has done using a rare and difficult material may turn out to be a guide to how to do the same trick using a common and cheap material. This is not invariably so, but the possibility must be kept in mind before an estimate of atrocious cost is used to stop an otherwise promising research or product development project.

At this stage it is also important to consider the question of how the product can be manufactured; indeed, the process of product development may be as much or more of a process of manufacturing system development as of development of the product itself.

For these reasons, product development has to be seen as a rather systematic and systematically planned procedure. In the earlier stages of research and exploratory development, producing the result is the key question. The question of replication of that result in large numbers and how one would go about it may be interesting, but it is certainly secondary to making the basic thing happen in the first place. In product development, however, the nature of the product, how to produce it, what it costs, and who will buy it become the essence of the matter. This is the stage in which business requirements and questions enter in a full-blown way.

Determination of Risk—The question of risk now becomes important. There are both technical and business risks to be considered.

One must consider the risk that the product envisioned may, in fact, not be technologically realistic. Even though all the facts and knowledge are believed to be known and there has already been an engineering demonstration of the feasibility of the concept, it may still turn out that there is missing knowledge that may or may not be available from further research, the demonstration of feasibility cannot be translated into a really reliable product, or the product can be produced only by manufacturing techniques that are not feasible, either for the required volume production or within the necessary cost range.

This is a stage in which testing, redevelopment, further testing, and further redevelopment become important, and where even the costs of this rigorous process begin to be significant in terms of the effect of their amortization on final product costs. The technological problem of testing may be significant, not only in terms of the dollar costs to accomplish it and the risks and uncertainties of actually getting the product, but also in terms of the time that must be allocated for the necessary testing for reasonable certainty of the result in terms of knowing product characteristics correctly.

At this point in the process the costs are sufficient and the product planning must be far enough advanced so that one begins to think in terms of firm schedules for bringing a product to market. This kind of "schedule thinking" is required even before the certainty of testing has demonstrated the availability of a real product, certainly before mass production has begun and has demonstrated the feasibility of the product and the process to produce it. Accordingly, at this point there is a significant gambling risk for any investment, recognizing an uncertain possibility of success.

At this stage there must be introduced significant investment risk capital or venture capital, either from within the firm, by borrowing from money or other markets, or by borrowing from venture capitalists who are in the business of placing such bets. The mere virtue of an idea at this point is not likely to be sufficient to convince everyone that a sensible investment should be made. There are complicated problems of estimating whether there indeed will be a market for the magic product, if it is indeed possible to produce it, along with estimates of detailed costs, and how long one can manage to carry on development and testing before passing beyond the right moment for market introduction.

This is the part of the process that tests the fundamental attitudes of the firm and the money markets upon which it depends. It is clear that some American innovation has been destroyed at this phase, because short-term possibilities of turning money over in other than product development and marketing are competing with the possibility of long-term, and perhaps larger but riskier, gains to be made by building a business and a market on a new product. While it is certainly necessary for a business to do well enough in the short term to survive into the longer future, it is not always clear that the long-term profitability is to be maximized by taking the short-term gains.

Current business practices of fancifully pessimistic discounting sometimes have led to a short circuiting of the innovative process and a concentration on money manipulation rather than product creation. There seems to have developed a general habit of considering only short-term payoffs to be worthwhile while discounting all long-term future possibilities, in spite of a strong track record of innovative companies which have spent as long as decades building product possibilities which then became major market opportunities.

These problems of risk are further complicated by the changing time scales for market development and competition. In previous eras, U.S. business operated with the confidence that it was sufficiently technologically advanced with respect to other parts of the world so that it could work out its technology at leisure in the belief that others did not have the fundamental capability to catch up even when they saw the product. The entire process was then considerably more relaxed. In the past twenty years, however, the ability of many parts of the world to develop high-technology products, based upon their own fundamental technological understanding and ability to either copy or independently develop what they have seen, has added considerable pressure to shorten the time required for product development.

Currently, rather complex difficulties, previously much simpler, surround even questions of proprietary rights, secrecy, and the ability to protect ideas with patents. Not only is it the case that many aspects of modern technology development are difficult or impossible to patent or copyright, e.g., software, genetic change, and many computer related ideas, it is also the case that patent protection, when achieved, is likely to be so specific that a sophisticated outsider, seeing the result and examining the patent, can invent a new way to do the task which totally by-passes the patent.

First to Market or Second—It is necessary to consider the entire question of whether one wants to be first to market, paying for the research and development that leads to the product, or whether one might benefit by lagging slightly and putting the effort into developing a second generation when it has become clear that the first generation of products is a market success. There appear to be cases in which the ability to come second to market very rapidly may have short-term business advantages for particular products. However, an understanding of the knowledge and technology is still essential.

From the point of view of the overall capability of the firm, or, in the long term, the nation, such decisions can result in the loss of innovative and technological capability. It is difficult to believe that one can repeatedly be second to market over a long period of time—never be a primary innovator—and still retain the necessary edge to succeed in high-technology innovation, or even "back-engineering" or "redevelopment" innovation.

It must be noted, however, that it is sometimes the first development that contains the bulk of the mistakes, and the second developer can perhaps profit by avoiding some of the first difficulties. Even so, it does appear that in the long run the knowledge necessary to have the capability to make the choice between being first innovator or first copier needs to be maintained if the innovative process in the firm or the nation is to be healthy. This aspect of the matter deserves considerable further study.

One area of this question that needs examination is the effect on the reputation of the firm and its products achieved by innovation or by copying, both on the firm's position in the market when it brings a new product into public view and on the motivations and morale of people within the firm who are responsible for the innovative processes that take place. Certainly a reputation for newness, advanced product concept, and technology is bound up with a reputation for quality and reliability and thus with the whole matter of what price the public will pay for whose product. Indeed, the whole matter of what product the public perceives coming to market (and perceives itself buying) is bound up with the question of the innovative process in an important way and is not really well understood. For example, when the Polaroid camera and film were introduced, the original market intention apparently was for snapshots and personal photography. However, the product appears to have been carried through its initial years by the fact that it was the best available technique at the time for recording what happened on an oscilloscope face; the early film market was dominated by electrical and electronic engineers taking data. As other means improved and the public market became established, this product innovation peculiarity changed. Similarly, the original market for microcomputers was an engineering market, shifted to a device-control market, and is shifting now to a public consumer goods market. It is probably difficult to plan these shifts; a certain amount of luck may be required.

THE OVERALL PROCESS OF INNOVATION

It is clear that what has been described as the innovative process—beginning with research, proceeding into exploratory or advanced development, and continuing into product development, manufacture, and marketing—is a complex process with a variety of aspects. The numerous feedback loops have been mentioned above, but it is important to emphasize that the social nature of the various parts of the process may be very different. The untidy, artistic, inspirational nature of research and early development has been referred to, but it must be noted equally that, while "creative intuition" is important in the management of manufacturing and in the problems of risk-taking and business analysis, a good deal more tidiness is required in these aspects. Systematic methods and tidiness are also important to the technical areas of careful reliability testing and quality analysis and assurance. While there are artistic aspects to all parts of the innovation process, the systematic aspects are much more emphasized at the product manufacturing and marketing end, and the inspirational aspects emphasized more at the research and development end.

TECHNOLOGY TRANSFER AND INNOVATION

The problem of technology transfer, matching different kinds of technological and business cultures as one proceeds through the innovative process, is complex. Basic scientific knowledge must be transferred by the research people to the development engineers, who differ somewhat from them and also from the product engineers who are next in line; the manufacturing engineers differ from all the preceding in their interests and responsibilities. Finally, business and financial people come from a still different culture and are likely to look for hard demonstrations in areas where scientists and engineers are accustomed to considerable ambiguity in terms of analytical and experimental proof of results.

Technology transfer, in any case, cannot be achieved merely by passing specifications, patents, or research reports to other people. Consenting adults must be engaged in the process, and a great deal of communication and mutual accommodation between the various parties must take place before a real understanding of both possibilities and requirements can be achieved. It is characteristic of the process that neither the product engineering nor marketing people have a very clear perception of what the technology can really do for a possible product. They, as the customers for technology, are always wrong. The research and early development people, on the other hand, have a tendency to see good products in every technological triumph and to try to convert each laboratory success into a business result. The mismatches in communication arising from these different attitudes and expectations can easily destroy the possibility of bringing the technology and product ideas together to a successful conclusion.

The transfer process from laboratory to market appears to work satisfactorily only when rather intimate dialogues among the various parties can be achieved. Frequently a period in which individuals actually transfer from one part of the process to assist in establishing another part on a firm basis may be necessary. For example, research people and engineers from early development may need to participate, or at least be available for detailed discussion, in portions of the product engineering process.

The rigid application of cost-accounting methods that were designed for different kinds of products and other situations can be death to a successful development, in that it may prevent easy communication (travel is expensive), misallocate costs and distort incentives, and hamper or prevent the development and technology transfer process since accountants do not fully understand this process. (This problem will be discussed further below.)

Because of the necessity for dialogue and feedback, a rather complicated process, we must turn our attention to the question of the organization of the firm for innovation.

The Organization

> *... the life of man, solitary, poor, nasty, brutish, and short.*
> Hobbes, *Leviathan*

> *Experts should be "on tap but not on top."*
> Attributed to A. E. by L. Gulick

The principal point to be made about organization for innovation is that bureaucratic matrices cannot be diagonalized; that is to say, there is no perfect organization for any particular purpose. Any hierarchical system will raise divisions between activities that need to communicate. Functional organizations tend to make tight development projects difficult, while project organizations tend to stifle the development of long-range deep technological competence in fields that may later be needed for other projects. This inherent difficulty frequently leads to cycles of reorganization, as one organizational system after another is tried and proved inefficient. It leads also to various kinds of matrix organization in which individual development or business centers may simultaneously be partly answerable to a project and partly answerable to a functional leader responsible for maintaining a basic technical competence and capability.

This may be one reason why small business units, and commonly small firms, can proceed most rapidly and effectively on many new innovations: they are small enough so that organizational requirements are minimized, and they can operate with fairly complete informal communication among all parties. When the organization grows beyond a certain size, this rather simple and unorganized arrangement becomes too unwieldy since the number of communication links grows in a combinatorial and, hence, exponential way.

The introduction of hierarchical organization to maintain a sorting of roles, tasks, and organizational control ruptures many of the seminal communication links that operate in simpler organizations and poses the problem of how to keep intact and healthy the communication lines required for the all-important technology transfer of innovation. Matters of centralization and decentralization of research and development in the firm (the establishment of central laboratories as opposed to laboratories and development organizations spread among manufacturing divisions but linked to the central research and development organization) should not be seen as solutions to a universal problem. Rather they should be viewed as attempts to introduce organizational devices that produce a good compromise for the preservation of the various communication links between research and early development and with product development and business aspects. Because what is to be preserved is a set of communication links which pass information and transfer technology in various ways, not any particular set of organizational niceties, frequently bureaucratic arrangements in the firm or in government, proceeding from standard hierarchical or organizational models, are more destructive than facilitating of the basic innovative process.

The problem has been made chronically worse over the past twenty years as firms are increasingly organized in terms of financial and accounting arrangements rather than product, marketing, or innovation arrangements. This is often true even when the parent organization is structured by product or R&D and manufacturing lines.

The problem arises because the details of accounting and bureaucratic systems may result in communication difficulties, as well as in assignment of incentives which interfere with the communications important to the innovative process. For example, the establishment of separate cost centers may make it difficult to have the adjustment of engineering and business requirements and the communications needed in an innovative development; this is particularly true if these are also profit incentive centers. Such difficulty results, for example, if it is necessary for one party to increase costs while another party can decrease costs, even though the sum of the two is lowered, and the result is to the benefit of the total corporation.

The problem also arises because of organizational rules for communication, the styles of various executives, and sometimes questions of legal arrangements among various parts of a firm which are intended to satisfy other kinds of rules and criteria.

Thus the question of business, accounting, and organizational systems in innovation must be seen from the point of view of their effects on communications and the ability to take a large-scale system design attitude to the product, the process for producing it, and its eventual marketing. Systems that systematize and segregate the various parts too well may distort the incentives for, and destroy the possibility of, real innovation.

This situation is complicated by the fact that recent attitudes have emphasized financial results to the point that there has been a loss of interest in the product itself. It has become a symbol of profitability rather than the essential element. Thus the operation of the firm can be seen increasingly in terms of cash flow, cost control, and accounting procedures, along with the systems of organizational control, as means by which there is control of a financial empire—a situation in which the idea of innovation gets lost along with the idea of the primacy of the product and its marketing.

It has been the conventional practice in recent years for accounting and business people to be on top, while the innovators are merely on tap and judged by financial criteria which may or may not be appropriate to the system characterizing the innovation under discussion. It is not clear that if the innovators were always on top, the result would not be a complete distortion of the cost of the financial possibilities of the innovation. The point really is that the purpose of a business is to make a profit by producing and selling products of some sort, and recent business practices seem frequently to have lost the central impetus of this idea. Because so many of the recent accounting and business practices are hedged about by external requirements of government, legal standards of accounting, business practice, and law, it is necessary to consider these external factors by themselves.

External Factors and Sectoral Relationships

> *My object all sublime*
> *I shall achieve in time—*
> *To let the punishment fit the crime—*
> *The punishment fit the crime. . . .*
> *On a cloth untrue,*
> *With a twisted cue*
> *And elliptical billiard balls!*
> W. S. Gilbert, *The Mikado*

Several aspects of the social responsibility and the ethics of the firm in bringing innovations to market have become institu-

tionalized in recent years in ways that have probably been deleterious to innovative drives in industry, although in some respects
they have succeeded in bringing protection of the public to a
new and useful degree of awareness and accomplishment. Whether
they have produced this result in a way which had to have the
adverse consequences it did, and whether this is essential to
producing this result, is open to considerable question.

Already referred to are the questions of accounting standards
and the application of a variety of tax standards which produce
accounting arrangements that may or may not foster the kinds
and nature of internal communications that are important. The
computer has made multiple accounting systems within a firm
possible, accounting legally and correctly for tax purposes, while
also accounting for costs in such a way that incentives can be
established that have the right impetus toward innovation. However, few if any firms appear to have seized the opportunity to
treat internal matters of incentive and understanding of costs on
a rational in-house basis while having separate accounting systems
for fulfilling tax and other accounting standards established to do
somewhat different things. The two are no more than mathematical transformations of each other.

An example of this is the extreme artificiality of depreciation
and amortization systems in tax and accounting standards. It appears that just at the time when it might be possible to account for
depreciation using the actual life of various pieces of equipment
as a sensible way of dealing with the problem, we have moved
even further into artificial standards for depreciation and have
tied tax incentives to depreciation periods which may not have
much to do with the actual use of hardware. The same has to be
said about the basis for taxing inventory gains, as in the introduction of first-in-first-out (FIFO) and last-in-first-out (LIFO), just at
the time when real accounting is simultaneously possible, cheap,
and very likely the best management control system.

The point is not whether these have useful aspects as incentives
for action by the firm, it is merely that their mindless translation
into internal accounting standards and financial incentives may
destroy many of the things that they were intended to foster. It
appears that broader analysis and greater thought about this
problem might be advantageous for the improvement of U.S.
innovation. The question is less one of subsidy and tax burdens

by sectors than it is one of the internal technology of translation for the firm's own self-understanding and the proper construction of its internal and external incentives.

PRODUCT LIABILITY

A second important problem is the growing body of product liability law and practice and its increasing form as a legal adversary system which has become more gladiatorial than judicial.

There is no question that a reasonable liability to the public of an innovator and manufacturer for the consequences of a product is sensible. However, this is increasingly being translated into the assumption that everything which was produced must be always safe and risk free under all circumstances, no matter how used and no matter whether the unanticipated harm may be of insignificant frequency.

It is extremely difficult even to deduce what the standards are for regulating estimation of the existence of a product defect, whether systematic or random, in either manufacture or design. The criteria are unclear. Increasingly the establishment of safety and environmental regulations proceeds on an assumption that it is possible to deal with the natural world without any risk of random events or human failure, and in a situation in which complete safety is always preserved with no untoward or unexpected consequences from any cause.

The fact that this assumption is almost certainly a violation of natural law and the way in which the universe is constructed seems of little interest to those asserting such views. This seems to occur because the basic facts about probability and the manner by which things are really designed and manufactured are quite unknown to those who have not been educated in any way in engineering and science. The legal profession, regulatory and judicial, appears not only to be proceeding in considerable ignorance of science, but is, in fact, developing a set of rules of evidence that is quite independent of, and in contradiction to, what is known about natural events. Legislation is frequently even further from any scientific relevance.

Unfortunately the counterbattle of industry against this difficulty, as well as some of the problems of regulation to be mentioned below, has not always been rational or entirely honest throughout. The legal defenses of corporations have frequently been based on countering irrationality with irrationality; the whole process becomes a jousting contest rather than a proceeding seeking justice. A battle of foolishness has ensued.

Thus the questions are not whether there should be safety and environmental regulations, or even whether these regulations should be both stringent and stimulative of important technical advances for the protection of public and environment. Rather, they are whether it will be possible to establish such regulations on the basis of reasonable understanding of what is known, knowable, and unknowable, and what classes of risks may be acceptable for the public and in accord with nature. The process of assuring minimal risk seems to be proceeding increasingly by a random establishment of things to be controlled, without any particular examination of the hierarchy of dangers to which anyone or anything is exposed. Perhaps it is impossible to do better while preserving everyone's rights in the context of a democratic society, but it does seem that a more reasonable set of approaches should be devised.

The importance of this comment on personal risk and product liability within the context of this chapter is simply that the exposure of innovative possibilities to the fear of random regulation of supposed consequences has what is nowadays called a "chilling effect" on the interest of firms and of individuals within firms to produce innovations. It is already the case that the ability to think through possible safety improvements is hampered by the fact that there are regulations which would make them improper, and, therefore, considerable nervousness exists about either the political or legal possibilities of changing those regulations without getting embroiled in intolerable political and legal situations. We are sometimes telling the goose not to bother with the golden egg because there is a faint suspicion that the shell might possibly be bad for health or environment. It is easy to "pooh-pooh" this possibility and regard its remoteness as a defense of doing things in an unethical or irresponsible way, but that does not make the fact of its effect on innovative people any less real. Innovators are beginning to feel as though the apparent attitude of those who do represent, or claim to represent, the public is something along the lines of "go see what Johnny is doing and tell him to stop."

GOVERNMENT, INDUSTRY, AND INNOVATION

Aside from questions of regulation and the like, the role of the government in sponsoring research and development can be important in its effect on the possibility of industrial innovation. It is particularly important that the government continue to take a long-term view of the sponsorship of long-term research and development. Most firms simply cannot take a very long view of their responsibility for knowledge generation for future innovation; this is a task in which the commonality of interests of the country must engage. The idea that everything undertaken without an immediate market view is unlikely to be worthwhile is merely a piece of foolishness in the face of both history and reasonable economic analysis. Markets choose things in terms of a hierarchy of immediate benefits and are notoriously bad at long-term thinking.

Neither does rigid government planning have a good track record. However, a pluralistic means of letting a large, thoughtful community work on long-term research and technology problems does seem to have been productive in the stimulation of innovation and new production in the U.S. Certainly the world production of new technology owes a great deal to the long-term funding and support by government of basic scientific and engineering research, and it even owes a good deal to the support of the development of that research into technology available for product development. One has only to look at the development of the computer and everything that it is bringing to us, the development of radio and microwave communications, and related items to understand the meaning of this assertion.

It is considerably less certain that government should engage in demonstrating that technologies that are clearly possible in principle can be turned into pilot products. This verges on the subsidization of individual industries or firms and may be so strong a central control of what product is produced as to derail the possibility of broader innovative efforts. Perhaps some means can be found to involve the totality of an industry along with government in contributing to the development of the demonstration of a basic technology and leave the individual firms to construct the means for production. This seems to be closer to the pattern that the Japanese have been employing quite successfully than it is to any U.S. attempts involving government sponsorship of demonstration projects. Before this can be successful, a rethinking of the nature of necessary antitrust protections is in order.

Conclusion

> *When you're lying awake with a dismal headache*
> *and, repose is tabooed by anxiety,*
> **W. S. Gilbert,** *Iolanthe*

Innovation is a complex and fragile process involving elaborate communication among a variety of cultures and different kinds of people. It can easily be destroyed on a Procrustean bed of bureaucratic arrangements and "management." This involves the paradox of the self-defeating solution: systematic management can in fact prevent the good management of innovation. Innovation must be managed in terms of the requirements of the process and not as an incident of the management of other aspects of the firm. To do otherwise is to mistake the scaffolding for the structure.

In this sense, we need the development of new business and social regulation concepts and technology, for it appears that the means for management, particularly of innovation—public or private—have not progressed as rapidly in the technological sense as our ability to develop new technologies and apply them to both old and new problems. It is not a question of determining whether the business managers or the innovators are to be on top; it is instead a question of constructing new means for all cultures to work together toward common ends.

During World War II and during the explosion of American innovation thereafter, we found those means, but in excess of enthusiasm for formal managerial systems we appear to have forgotten how to use them. Perhaps we need to revisit and relearn our past successes so as to bring new technology, both social and technical, to the future innovative process.

Willard Marcy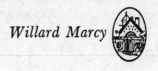

4

Enhancements and Impediments in the Innovation Process

Innovation

Technological innovation is the transformation of new concepts into needed or desired products or processes not previously available. The transformation is accomplished by a complex process involving complicated organizational structures, a dynamic process subjected to and modified by frequent changes resulting from technological, environmental, and societal influences. Logic and emotion play key roles; science and art are importantly involved.

NATURE OF THE INNOVATION PROCESS

Erik A. Haeffner of the Institut for Innovationstechnik in Sweden presented a detailed and provocative analysis of the in-

WILLARD MARCY *is president of Applied Research & Development of University Science (ARDUS), a subsidiary of the Drug Science Foundation. Dr. Marcy was previously with Amstar Corporation and vice president of the Research Corporation. Among several other professional activities, he is an active member of the American Chemical Society and the American Institute of Chemists and serves on the board of trustees of The Chemists' Club in New York. Dr. Marcy has written numerous articles and papers on innovation for national and international publications.*

novation process in the March/April 1973 issue of *Technology Review*. He points out that during the first sixty years of the twentieth century, students of the innovation process almost universally assumed that basic scientific research would lead automatically to technical progress, that development of research results would lead directly to new products and processes, and that economic growth would follow. Unfortunately, several studies designed to show a correlation between the results of basic research and the number of innovations in a given industry indicated that a major impetus for industrial progress comes from inventions which are not a result of basic research.

While scientific research increases knowledge and provides an essential base from which new ideas and inventions can flow, it is now generally accepted that the major motivating factor in the innovation process is the condition of the marketplace. The existing and expected future economic atmosphere of a given industry largely determines whether inventions are created and developed. A world ready for minicomputers and microcomputers has provided the impetus for innovations in microchip and integrated circuit design, fabrication, and methods of use. An invasion of less expensive, more reliable, and more efficient automobiles from overseas has stimulated domestic manufacturers to emulate and improve on innovative production and quality control techniques used by foreign car manufacturers. Innovations such as these, however, cannot occur unless the scientific information base has already been developed. The market may be ready for an innovative advance but may have to wait for science to produce the required knowledge to develop new products.

Innovation is Dynamic—A further complication arises from the dynamic nature of the innovation process. As scientific knowledge is gained and understanding of market requirements increases, changes in the directions of research, development, production, and marketing are required. The people involved also change their perspectives, mature, lose interest, and are brought in when new problems arise. Every change affects the course of the innovation process in varying ways and somehow must be taken into account.

Dynamism occurs not only in tangible and mechanical areas, but also in the realms of the emotions and spirit. A sudden illness or incompatibility of personalities can cause real disruptions. Personal relations and managerial styles are very important and can change quickly when people are working together on a common problem. Differences of opinion, perceived slights, misunderstandings, and even moral and ethical problems all alter over time and need to be resolved for success to occur.

Logic and Creativity in Innovation—As much as one might like to think the use of rational, scientific logic will produce inventions, innovation in most cases is the result of a creative process in which an invention is made, followed by an often long, tedious, roundabout development which leads to the final product. Rational research frequently is used in the development process to aid in solving problems that arise en route, but such use is secondary to creative thinking, although often financially important.

Creativity in innovations occurs during the attempt to cope with many factors not normally considered important by scientists and other technically trained individuals. Chance contacts, serendipitous events, and unpremeditated discoveries can change drastically the effectiveness of the innovation process and the speed with which it is accomplished.

The Role of Communication—Communication plays a major role in the innovation process. For optimum and timely success all the players need to understand each other's roles and coordinate their efforts just as in a theatrical production. Here is where the art comes in. The innovator has the task of melding both intellectual ability and facts to produce new and coherent concepts and marketable products. If members of the team will not or cannot communicate their knowledge and ideas to one another, a common goal can be reached only with difficulty, in an untimely fashion, or not at all.

The Management of Innovation

Innovation starts with an idea. The idea is ultimately embodied in a device, a substantive material, or a process for accomplishing a purpose. The embodiment of the idea requires an interdependence of skilled people—inventors, engineers, mechanics, production experts, financial managers, marketing experts, and salespeople. Thus, innovation requires managing ideas, material, machines, and people.

Good management is even more important in the innovation process than it is in an established corporation. The well-known authority on business management Peter F. Drucker has predicted that the very heart of management involves entrepreneurial innovation and that social and technological frontiers will challenge the manager of the future. Instead of operating in a closed system, as has often been the norm in the past, managers today must cope with constant change while still maintaining continuity with the past—a situation which exists in any innovative atmosphere. Managers of innovation must rely primarily on subjective judgments; a science of innovation management has yet to be developed, and the management standards and procedures developed in the past are not adequate to the present-day requirements of new ventures.

QUALITIES OF INNOVATION MANAGERS

Leaders with special talents, capabilities, and knowledge are required to manage the innovation process. Such leaders are rare, and successful ones stand out from other managers. Why they are successful has been the subject of hundreds of studies over the years, but their secrets still remain hidden even to successful entrepreneurs themselves.

Successful innovation managers combine the talents of inventors, entrepreneurs, and businessmen and women; it is the rare individual who can adequately fill all these roles alone. Therefore, a number of individuals are normally involved in any specific innovation. But each of them resembles each other in certain general ways. Among other things, each must be intelligent, creative, energetic, and have a high degree of integrity with focused goals. Determination, persistence, and singleness of purpose, combined with flexibility, are also key characteristics. Innovation managers must have all of these and, in addition, an acute sense of time. They must know instinctively when to take risks and when to pull back, when to be aggressive and when to compromise, when to be tough and when to be tolerant, how to motivate colleagues and associates, what customers want and how to satisfy their desires, and how realistic but imaginative financing can be obtained and utilized; they must also be able to accomplish all these things within legal, ethical, and societal bounds.

Innovation managers start small and build; they may even be inventors. But in any event, they think they know about inventing, developing an idea, and building a business. They know that it is not enough to organize an effective team of scientists and engineers and expect markets to materialize like magic. They also know that even though a market may be there, the creative people who conceive of a product initially may not be able to undertake its commercialization successfully.

THE INVENTION

The innovation process starts with the inventor. While another person may have better knowledge of market needs and may communicate these to the inventor, it is inventors who originate specific new products or processes with detailed, qualitative functions. Inventors must be insulated from negative influences while they are inventing intensively, and they must have adequate and patient financial support. Risk of failure is substantial.

THE DEVELOPMENT STAGE

When inventors begin to develop a product or process idea, and often even earlier, they need the help of an enthusiastic champion, a person who cannot only give encouragement but also can be realistic about the next steps to be taken—and often actually undertake these steps on behalf of the inventor. Usually this is where innovation managers first enter the picture; they must provide patience, faith, and utter confidence in the worth of the inventor's findings and their own ability to carry through to the marketplace. They must know how to accomplish things through other people. Together, the inventor and the innovation manager can carry the innovation through its developmental phases and into small-scale production. Once this is accomplished, risk of failure has been greatly reduced.

THE GROWTH STAGE

As long as the market is small and the operations remain relatively simple and straightforward, these two individuals can handle almost any situation. However, with company growth, complexities develop requiring a higher degree of organizational ability, and knowledge of finance, manufacturing, marketing, sales, and personal relations usually not found in people who are inventive or have strong entrepreneurial bent. While creative thinking is an asset, creativity itself is less important at this point, since, now, organizational procedures based on sound business principles become essential. Here is where business experience is needed to decide whether or not market needs require expansion and how to develop the organizational capability to accomplish such an expansion. Innovation managers must find financing, obtain other managers, control operations, and deal with governmental and societal influences. Risk of failure at this time is minimized. During this stage inventors and scientists generally have great difficulty in managing their innovation. They are too involved in their brainchildren—trying for perfection beyond consumer needs, attempting to run all areas of the business themselves, being unable to bring in additional people with specialized expertise, or having difficulty delegating authority to others as the business expands.

THE MATURE BUSINESS

As the innovation process enters the mature stage, it usually provides only one or perhaps only a few products or services. Favorable customer response requires organized expansion—sometimes very rapidly, as has happened recently in the home computer field. Successful managers of innovation must be able to anticipate expansion or changes in production facilities and marketing capabilities, needs for financial support, personnel, and raw materials. And they must be ready to modify the organizational structure as necessary to maintain control of all aspects of the operation.

Manufacturing and Production—Manufacturing and production concern the manipulation of materials and devices using human intercession. Managers of manufacturing and production must have a good historical perspective not only on the specific industry and its markets, but also on how new products and processes must be made to satisfy new market needs in the industry. Engineering know-how, together with the ability to listen to and adjust to relevant feedback from marketing and sales experts, is essential. In new ventures, creativity and flexibility tempered by in-depth knowledge of science and technology are needed in order to produce cost-competitive new products.

Marketing Products of Innovation—Marketing and selling new venture products require time, much patience, and knowledge of the right people to approach. Early awareness by potential users of new products usually depends on external sources, such as reports in scientific and technical publications, advertising, and vendor contacts. Evaluation of new products leading to adoption in the market depends to a great extent on personal communication with technical- and management-level people within the users' firms. At each stage in the marketing process the availability, the quality, and the cost of the new products have a great impact on acceptability. The market for which new products are designed requires close and detailed study to determine the role of competitive products, how firmly established these may be, and how long it will take to obtain a profitable return after marketing begins.

Innovation Depends on Optimum Financing—Knowing how much financing is needed and where to obtain it for each phase of the innovation process is crucial. Too much or too little financial support can kill a venture even with good prospects of commercial success. Obtaining the wrong kind of financing from the wrong source is also deadly. Loss of control of a new venture can easily occur unless care is taken to deal with sympathetic financiers with a genuine interest in the overall success of the innovation.

Organizational Structure and Innovation—Within the organization itself the managers must decide whether the optimum results will be obtained under a hierarchical structure, a decentralized organization, some combination of both, or an entirely different organizational structure. The extent to which a bureaucratic system is used must be decided as opposed to a looser managerial procedure which might enhance creativity and productivity. After a study of successfully managed businesses, Thomas Peters and Robert Waterman, Jr., in their book *In Search of Excellence,* feel that the coming epoch of organizational thought will emphasize informality, individual entrepreneurship, and evolution. Recently, successful companies seem to emphasize flexibility in management, sometimes using entrepreneurs as product champions and, at other times, using a tough, autocratic approach, whichever seems appropriate under the circumstances. Past managerial wisdom, on the other hand, has emphasized the rise of military-like organizations which allow only limited ways to organize and solve problems.

GOOD PLANNING IS ESSENTIAL FOR A SUCCESSFUL INNOVATION

Planning is essential for the success of the innovation process. Even the creative acts of the inventor should be planned to some degree, being careful, however, not to go so far as to extinguish the creative spark. Planning reduces uncertainties in the risks; helps to reduce the effect of or eliminate competitive or externally generated surprises; enables one to distinguish among alternatives in the use of time, effort, and resources; and defines and redefines market potential. Planning also has a dynamic component; it should be done frequently as the venture matures.

Intellectual Aspect of Innovation—Successful management of innovation by innovation managers requires that they grow with the business. They must understand and take into account new and important factors related to expansion of plant and personnel and to the plant's managerial and financial requirements. They must turn their attention to optimizing resources and profits. They must develop strategic plans for the future and need to understand, manage, and control interrelations among these factors.

Societal and Political Content of Innovation—Management of innovation has a large societal and political content. Factors in these areas bring the need for applying ethical, moral, and legal judgments to the management of an enterprise. Successful management takes into account consumer needs; customer satisfaction; environmental impact, both favorable and unfavorable; and relations with competitors, community, and, in some cases, international entities. All legal restrictions, regulatory agency requirements, and health and safety regulations have to be understood and managed.

Enhancing Innovation

INCREASING SCIENTIFIC KNOWLEDGE

Practically all technological innovations rest on a scientific base. Even innovations in service industries ultimately are dependent on scientific principles. The advance of cable television rests on functioning space satellites; electronic banking requires computers; automobiles are built at lower cost by using robots. But until a basic knowledge of scientific principles has been acquired, innovation in a given industrial area is not possible. With such a base, however, innovative ideas proliferate.

Continual scientific research is essential to increase this fund of knowledge; both fundamental and applied research must be supported. Fundamental research is best performed primarily in an academic setting and is primarily supported by government agencies and public and private philanthropies. Debates take place annually relating to the magnitude of expenditures of tax generated public funds to be expended in support of fundamental research. Major support from this source in 1982 amounted to $4.6 billion, about 50 percent of which was provided by the National Institutes of Health (NIH) and 15 percent of which was provided by the National Science Foundation (NSF). Most of the total funding was for mission oriented research in the Department of Defense, the Department of Energy, the Department of Agriculture, and similar agencies. Funding of fundamental research by philanthropic organizations and industry is currently only a fraction of public funding, perhaps 10 to 12 percent. Industry research funding is mostly for applied research to develop or improve specific new products or processes.

Increasing the funds available for scientific research would be expected to enhance innovation. Such funding should come from all sources, not solely from public coffers. Since support of basic research is a long-range, high-risk activity, results may be decades in coming to fruition. This presents hard choices for these funding decisions relative to the financing of other important commercial societal needs.

DEVELOPING TECHNOLOGICAL EXPERTISE

The mere availability of scientific information is not itself adequate to ensure the design of marketable products or processes. The scientific information must be analyzed, dissected, rearranged, and resynthesized into forms that can be marketed profitably. The transformation of scientific information into useful products involves people with technological engineering, marketing, economic, and financial expertise and is generally referred to as "development."

Any program to enhance the rate of innovation and increase productivity thus requires a systems approach, one that integrates the whole gamut of activity involved in bringing ideas from initial concept to valuable products. The innovation process is much more complicated than a simple step-by-step evolution from the base of scientific knowledge. Scientific knowledge and technological development must proceed concurrently. They must interact with each other through feedback loops, information-gathering, and dissemination centers. Fueling such interactions is market demand and the likelihood that profits can be made from the sale of new and useful products and processes. However, it is important to recognize that completion of the innovation process will differ in a number of ways, depending on whether or not market needs, technologically, are being filled and/or scientific advances are being developed.

DEVELOPING AN INNOVATION MANAGEMENT TEAM

Many students of the innovation process rate highly the freedom of action encouraged in the American democratic atmosphere. They feel the inquiring mind and inventive spirit of the American citizenry coupled with the ready availability of venture capital and the presence of many entrepreneurially minded technologists, make it easier, compared with other countries, to start new companies in the United States. Bringing together the right combination of people to realize the inherent potential in this favorable situation is still an art and often depends on fortuitous circumstances rather than detailed planning based on scientific principles.

Better understanding of the innovation process and its interacting factors cannot but help to increase the likelihood of ultimate success. Inventors not only must understand available scientific knowledge, where to find it, and how to use it, but they must also understand what technology exists and is needed to develop their invention, where to obtain it, and how to use it. A dedicated champion with the proper understanding of both the market and the product to be marketed and an adequate source of capital which can be relied upon for years of scant return are absolutely essential to success. Financial investors need to understand the needs of both inventors and business managers to tap the scientific data base and to acquire and use needed technological resources in order to provide marketable products. If these individuals find difficulty in working together for common ends, the enterprise will fail regardless of how sound the base of scientific knowledge or how substantial the market may be.

UNDERSTANDING THE MARKETPLACE

Innovation managers must have a thorough understanding of the position of their products in the marketplace. Too often this understanding is limited or based on insufficient data. Avoidance of this seemingly obvious, fundamental error would greatly enhance the success of innovative ventures.

Innovative products and processes almost invariably push older ones out of the marketplace. The size, complexity, and pricing structure of the existing market must be analyzed, and some idea of where in the market the new products or processes fit must be determined. Timing of market entry, obsolescence of both old and new products, and geographical factors need to be taken into consideration. Knowledge of the nature, strength, and possible response to competition of competing companies is helpful, but it can only be obtained through speculative judgments based on best estimates and intuitive thinking. Early market testing is essential to help prevent gross mistakes, but these tests must not be relied on as an absolute gauge of consumer demand or acceptability. In those industries where patent protection is important, competitors' basic patent positions are necessary to discover and keep in mind.

Understanding the product life cycle concept is essential; a business built on a single product or process may follow the conventional profit life cycle and decline as the product matures, eventually expiring as market acceptance disappears. Successful businesses have a number of products or processes which continuously grow mature, and are replaced by new ones, producing an averaging of profits over extended periods of time.

PATENTS AND COPYRIGHTS

Patent rights ownership is very important in the chemical and pharmaceutical industries to protect newly marketed products from competitive pressures so that the costs of research and development can be recouped. Small businesses also benefit from patent rights ownership as a protection against large, predatory companies. Patents are less important for large companies and the mechanical, electrical, electronic, and service industries. Where patent coverage is essential, obtaining strong claims that can be enforced effectively against unlicensed competitors enhances the innovation process.

Copyrights ownership can play a major role for companies directing their efforts primarily toward marketing and other services. For example, recent court decisions have provided copyrights with a major role in the protection of computer programs from misuse by unauthorized parties.

GOOD COMMUNICATIONS ENHANCE INNOVATION

The success of the innovation process depends on good communications among people at all stages from the initial conceptualization through the life cycle of the commercialized products and processes. The means used to communicate and the people involved in it vary appreciably from stage to stage and change almost constantly.

At the outset, inventors obtain their innovative ideas from almost any source—the published scientific and technical literature, unpublished reports, personal contact, the media, or even daydreams. Their personal experiences are their frame of reference; they ask themselves the question, "What would happen if I did this or that?" Communication with others may be limited or nonexistent at this point; nor is it usually necessary until after their ideas have been tested in a laboratory or, perhaps, in a limited way in the marketplace.

As the innovation process develops, additional people become associated with the inventor and his or her early colleagues. One of these may play the role of an innovation manager. Communication now takes on a different, more complex aspect. New, broader sources of information are needed; communication lines are lengthened; a number of people, rather than just two or three, must be kept informed. Time becomes very important; analyses of available information must be made; bits and pieces of information must be integrated. Innovation managers and their associates communicate with outsiders—financial supporters, marketing analysts, legal counsel, and tax experts. Each of these outsider experts speaks a different language, and the inventors and/or entrepreneurs must learn, if they do not already have the facility, how to speak the same language. While some communication will be through written literature, as in earlier stages, most will be by personal contact in private meetings, both internal and external conferences, and at professional society gatherings. As more and different groups contribute to the activity, technology-interface problems arise, and personality conflicts begin to surface. These unfavorable situations must be resolved as soon as they are perceived in order to conserve momentum and minimize lost time.

Coordination of all efforts now becomes very important. Successful innovation projects involve much more communication, both internally among project personnel and externally with colleagues outside the project, than less successful projects. Outside communication of the successful groups includes contacts within their own specific discipline and others as well. Any action which promotes frequent contact within and among disciplinary groups improves research and development effectiveness.

As the communication network grows and expands, individuals who possess a special facility for communications emerge. Others naturally turn to these individuals for help in arranging contacts and obtaining information. These key people become "technological gatekeepers" channeling information flow in and out of the organization; they are especially important in large, geographically separated organizations. Generally speaking, such gatekeepers are high-technical performers and busy first-line supervisors, interested in a wide variety of outside activities. Management, entrepreneurial management especially, is well advised to identify these individuals and encourage and develop their capabilities.

When an innovation enters the commercial stage, good communication in all its aspects must be developed in all facets of the operation. This is particularly important at the time of first marketing. Customer reaction and its timely and accurate feedback to manufacturing and internal marketing departments are essential to make sure the products fill real market needs and produce satisfied customers. While communication needs change drastically in mature organizations from what they were in the early stages, sensitivity to these needs must be present throughout the innovation process.

The Role of Venture Capital

Adequate financing is essential throughout the entire innovation process. Unless financially sponsored by their firm or another interested party, inventors have great difficulty in accomplishing initial experimentation and testing of their ideas. Without sufficient funds, months and even years may be required just to produce a prototype or the first successful test. This situation is not all bad, however, since time, as well as money, is required for inventors to become aware of problems to be solved and means to be devised for their solutions. Backtracking and rethinking are normal and necessary to avoid falling into unforeseen traps. The availability of funding cannot substitute entirely for time during the early stages of innovation.

The nature, type, and amount of capital, as well as the philosophical and motivational attitude of the person or group furnishing it, change as the innovation process matures. In the initial stages minimal capital is required, prospects of eventual return are low, and the length of time before any return is received may be quite extensive. During this period opportunistic venture capitalists may find that it is expedient to loan money, with or without interest charges, in return for a stake in the enterprise. They must be patient and be prepared to endure disappointments, as well as be willing to provide additional funds occasionally.

When products are developed and marketing begins, additional operating capital is required more frequently and in larger amounts. It is necessary to support manufacturing and marketing efforts of sufficient size over a period of time that is long enough to indicate acceptability of the new products. The risk to capitalists is reduced, but the amount of money at risk is substantially increased, perhaps as much as ten times that required during the initial research, development, and testing stages. In addition to loans, public sale of bonds, debentures, or stocks, it is frequently necessary to raise adequate capital to carry a new venture through this phase of growth. Raising capital in this manner, however, entails dilution of ownership since purchasers of stocks or bonds own equity in the company.

AVAILABILITY OF VENTURE CAPITAL

Until the 1960s and 1970s the use of venture capital was considered a black art surrounded by an incomprehensible mystique, and the venture capitalist was perceived as something of a high-stakes gambler. This situation has changed greatly in the last two or three decades, and today's venture capitalist is regarded as a rational businessman or businesswoman with a well-studied understanding of the venture capital process and an organized approach to funding new venture.

While venture capital appeared to dry up in the early 1970s, in actuality there has always been more than an ample supply for small businesses and new enterprises backed by talented people with good business ideas with only modest funding requirements. The apparent lack during the 1970s occurred primarily in financing the expansion of already existing businesses whose marketing of products was adversely affected by depressed economic conditions. In the 1980s this situation reversed as the general economy became more favorable. Venture capital is now being perceived as plentiful, especially for those companies in high technologies—computers, electronics, and bioengineering, for example.

FINDING VENTURE CAPITALISTS

A major problem inventors and entrepreneurs face is finding compatible venture capitalists with sufficient means to support a new venture for the necessary length of time. Venture capital may come from investment bankers, mutual funds, individuals, family trusts, insurance companies, pension funds, commercial banks, corporations, public and private venture capital companies, small-business investment companies, or private partnerships (including research and development tax shelters). Each of these sources has different objectives, motives, and methods of operation. A few are interested primarily in frontier research; others, only in high technology with high return (albeit at high risk); still others, merely in established companies with products already on the market. Some have interests only in certain areas, such as chemicals, petroleum, or heavy industries; and others are involved solely in marketing proven products or services.

In general, venture capitalists search for high-risk, high-payoff situations. They regard the desired return on their investments to be in the range of 500 to 2,000 percent. As a rule of thumb, a 300 percent return in four years or a 400 percent return in five years is acceptable. However, those who were patient enough to be involved at very early stages of an innovative venture expect even higher returns over longer periods of time.

Innovation managers seeking venture capital must evaluate their needs realistically. If they raise more capital than needed at any given time, they are in effect selling more ownership in their venture than needed. On the other hand, insufficient funding leads to potential disaster resulting from underestimating or understating their requirements. The successful innovation manager also knows that financial requirements generally follow a relatively smooth curve, whereas obtaining capital is usually a stepwise operation.

Impediments to Innovation

The innovation process is hindered by any number of unfavorable circumstances. Some arise from internal difficulties, but many more emerge from external sources unconnected with the specific innovation under development.

INTERNAL IMPEDIMENTS TO INNOVATION

Most internal impediments to innovation are unrelated to the technical merit of the invention itself, but, instead, they arise from either the inadequacies of the people involved or unforeseen external circumstances.

Inventor Attitudes—Many inventors are eccentric; they try to beat the laws of nature, have an obsession they cannot or will not drop, and are forever searching for the nonexistent pot of gold. Inventors of this type are doomed to failure, because they do not possess sufficient scientific and technological understanding, knowledge of the marketplace, or adequate financial and economic know-how for them to succeed. They may attract entrepreneurial or even venture capital attention for a brief time, but their inventions (almost invariably without merit) wither and die, taking their backers with them; the innovation process is never completed.

Resistance to Innovation—Innovation resistance is a very real problem, not only within an organization, but in the marketplace as well. Recognition of this fundamental characteristic of human nature and devising means to cope with it are hallmarks of the successful innovator. Overcoming resistance to change has been the subject of much study by sociologists and pragmatic trial-and-error experimentation by managers and marketing experts. Present thinking states that resistance primarily results in regard to social change, not technical change, and from perceived changes in human relationships that involve personal prestige, worth, and interactions. Communicating the need for change beforehand, discussing the possible results of the change with the people who will be affected, and providing examples of benefits from the change are all useful techniques to minimize such resistance.

Poor Management—A Denver Research Institute study of 200 innovations that failed after initial commercialization reports that poor management accounted for 23.5 percent of innovations cancelled, shelved, or inordinately delayed. Many management errors seem preventable. Over 33 percent of the management errors involved market factors which management could have anticipated. For example, one company, at great cost, developed a welding torch for repairing automobile bodies only to find that potential customers viewed the torch as a fire hazard. Almost 10 percent of the failed innovations resulted from lack of a market; approximately 7 percent were blocked by competition; about 5 percent ran into patent infringement problems or antitrust law violations.

General Georges F. Doriot of American Research and Development Corporation has provided an interesting summation of management errors found in that organization's experience with start-up companies. They include:

1. becoming too emotionally involved in an idea or individual;
2. excessive delays in foreseeing problems or applying corrective measures;
3. the inability of entrepreneurs to grow with the business;
4. the inability of technically trained managers to stay knowledgeable in their fields;

5. acquiring a bureaucratic structure too early;
6. a lack of foresight;
7. excessive belief in the product under development;
8. inadequate knowledge of competition and the marketplace;
9. pricing products too low at the start of marketing;
10. poor knowledge of costs, overhead, and inventories;
11. a lack of understanding of the difference between operating profitably and having a profitably growing, competitive enterprise;
12. the premature breakup of the original team or, conversely, too great a loyalty to the original team; and
13. a greater interest in personal return than in building a viable enterprise.

From these limited examples of poor management, it is apparent that managers could save good innovations by asking the right questions at the right times.

EXTERNAL IMPEDIMENTS TO INNOVATION

Innovations undertaken without due regard to economic, environmental, and societal factors are in peril from the start. For the most part, economic influences arise from outside the immediate venture. These include such items as inflation, general economic recession, unforeseen political situations, or public opinion. These factors must be considered at the very beginning of innovations by inventors, entrepreneurs, and investors. As the innovation process proceeds, reevaluation of these factors must be undertaken frequently.

Public Opinion—Currently, public focus is on the contributions science and technology can make in the solution of broader societal problems. Critical changes have recently occurred in the international environment affecting world trade conditions and the availability of energy and raw materials. While some people believe these new focuses are cyclical and that profound changes will not occur in the future, the majority believe that fundamental changes in our society and economy will take place over the next few decades. Extrapolating from the extraordinary changes that have already occurred in the electronics industry (embodied in the swift changes being brought by the proliferation of the use of computers) and from the unpredictable, imminent, and profound effects that the infant industry of biotechnology is beginning to produce, it is safe to predict that the choices made by today's society will determine the society of tomorrow. These choices must be made with as much knowledge and moral integrity as can be brought to bear. The mere marketing of products for profit must be tempered with consideration of the larger and long-range consequences on society as a whole.

Government Involvement in Innovation—New and more restrictive laws and regulations are being promulgated in an effort to maximize benefits to the public. However, much of this activity has been reactionary rather than progressive, costly and inhibitory, and so far often limiting the optimum societal use of science and technology. To ensure a proper balance between risk and benefit will require much more study and discussion, as well as action in broader societal terms. Private sector priorities and judgments must be used as a major element in planning economic growth with the full realization by public figures that use of public funds may create private profit in the process of accomplishing national objectives.

During the 1970s and the 1980s, a piecemeal approach was taken by various public and private sectors directed toward improving conditions or enhancing the motivation in different areas of the innovation process. Many of these efforts have been misdirected or treated symptoms rather than basic problems. For example, Congress has "reformed" the patent laws with the objective of making it easier for research universities and institutes to obtain patent ownership of inventions made with government funding. However, such organizations have no means for developing the patents they own and must either rely on some patent-service agency or independently enlist individuals or industrial companies to recognize market needs and develop products to meet those needs.

Much federal-level discussion has occurred relating to the establishment of cooperative technology centers, centers of academic excellence, and similar institutions to increase the collaborative efforts among universities, industry, and government. Problems immediately arise regarding which party will play the dominant role in program planning, allocation of resources, and management of these centers and as to how the results and benefits arising from the work done at such centers will be utilized. As a consequence of an inability to resolve these problems, substantive action to implement these suggested activities has not yet occurred. The government-industry-academic interface has been and is continuing to be an area of intense discussion and study. While it is conceded that more effective cooperation among these sectors would enhance innovation, appropriate means for accomplishing this purpose have so far eluded definition, except in a few special cases.

Recognizing the desirability of transferring government owned technology to the private sector, the National Technical Information Service (NTIS) has been charged with disseminating scientific information and patented inventions developed in government research laboratories. However, only a small budget has been allowed and a staff of less than a dozen people provided. The scientific concepts disseminated, for the most part, have essentially no relations to civilian market needs since they were developed primarily for government purposes such as defense and space exploration. The NTIS program has been limited to writing and publishing a huge volume of descriptive material which is made available only at a few selected locations. While its availability is made known through listings in journals such as the *Federal Register* and government procurement notices, such methods of information dissemination are almost totally ineffective even though accomplished at substantial cost.

EFFECT OF LEGAL CONSTRAINTS ON INNOVATION

A surfeit of international, national, state, and local laws and regulations impinges on the innovation process beginning at its earliest stages—even while accumulation of scientific and technological information is occurring. Awareness and observance of these statutes are essential, since they are designed ostensibly to aid innovation, to regulate how innovation is accomplished, or to inhibit or prevent abuses—all done in the name of the common good. Observance of these laws and regulations inevitably leads to bureaucratic procedures and large amounts of paperwork that entail extensive clerical and bookkeeping activities. Such peripheral consequences must be kept in mind and taken into account throughout the innovation process.

In the late 1970s and early 1980s, the federal government sought to ease the inhibitory nature of a number of these laws and regulations and to provide further incentives to innovation through new laws. Reform of the patent laws, favorable provisions in authorization and appropriation acts for public-grant agencies, and favorable tax treatment for research and development expenditures were put into effect. However, these are all of a piecemeal and relatively timid nature and, while helpful, a long way from having any major favorable influence on the innovation process. Innovative leaders take into account these legal aids, but they do not depend on them to help reach their goals.

Food and Drug and Similar Regulations—A special case of restraint on innovation in the chemical and pharmaceutical industries is posed by the Food and Drug Administration's regulations and those promulgated by other environment, health, and safety agencies. The need for such regulations is not questioned, but the excessively restrictive nature of some of the regulations, as well as the inappropriate way in which some regulators apply them, have increased the cost and lengthened the time necessary to introduce new and useful drugs. In addition, they have inhibited the scientific research directed toward the discovery of new chemical entities of therapeutic value.

Science and Technological Policy

Innovation is a major source of economic growth; it can help control inflation, create jobs, and achieve a more satisfactory balance of trade. It is the single most important contributor to productivity improvement. Properly managed, it can contribute significantly to the improvement of living standards. In this broad sense innovation embraces not only technological changes but also includes new methods of management, financing, marketing, and distribution.

The complex and dynamic process of innovation involves a number of main elements, universities and their scientific information base, inventors, entrepreneurs, businessmen and businesswomen, the public, and the government. Accordingly, it would seem desirable to find ways to improve and enhance the interrelationships among these elements for the benefit of all. Throughout the innovation process, large investments of time, effort, skill, and money are required. Such investment, although highly risky, must be applied at every stage with an inventor-entrepreneur-businessperson as the driving force.

Innovation is accelerated when businesses invest in new plants and equipment; conversely, new and advanced technologies create a large demand for capital investment. When demand for improved products occurs at the same time as innovative research produces new technologies, economic growth occurs. Thus general trends in the rate of capital investment relative to gross national product reflect the vigor of economic activity. In the United States in past years, changes in the national cash flow and investment in plant and equipment have moved closely together, but since the early 1970s, cash flow has exceeded capital spending by a much larger and continually widening margin, suggesting an accumulation of cash by American businesses.

A *Business Week* survey found that industrial leaders were reluctant to invest their cash reserves because of uncertainties about the course of inflation and federal wage, price, regulatory, and energy policies. In addition, stringent conflict-of-interest rules have kept the best people out of government and inhibited access to industry experts. Most meaningful of all has been the reduction of potential rewards that are perceived to result from undertaking high-risk capital investment.

Many people believe development of a comprehensive formal policy at the federal government level is necessary to combat or alleviate uncertainties and to encourage an increased flow of innovation. However, most thoughtful experts in both industry and the private sector believe that an overall government policy would be too rigid and too difficult to enforce meaningfully in view of constantly changing economic and societal requirements.

The Industry Advisory Committee to the Federal Domestic Policy Review on Industrial Innovation has recommended that the areas of highest priority for policy change lie in regulatory reform and provision of tax incentives. The regulatory reforms are recommended to include better assessment of cost-risk factors and to provide guidelines for taking optimum advantage of industry's capacity to satisfy the environmental health and safety needs of the public through innovation.

The committee feels a specific preplanned policy may be helpful in devising tax reductions designed to strengthen investment incentives for plant and equipment. However, any reduction in tax revenue poses a dilemma: such a reduction could cause federal deficit increases that lead to higher inflation rates. Congressional action in the early 1980s provided a few minor revisions in the tax laws designed to favor additional investment in innovation, but more creative thinking is still needed on this issue.

Improvement in the theoretical assumptions made by economic

planners in calculating revenue impacts of alternative tax policies is greatly needed. Economic planners and forecasters are hindered by a lack of understanding of the interplay between the numerous factors affecting their theories and the results of their predictions. Present-day models of the economic structure are simply not appropriate, in the opinion of many experts.

The United States is such a large and diversified country and so indoctrinated in democratic principles that central planning of a science and technology policy by a small, albeit possibly representative, group of either elected or appointed officials does not seem appropriate. Such an approach may be desirable for small countries with limited resources or for underdeveloped countries, but the examples of central planning in the large socialist countries do not inspire confidence that this mechanism will provide adequate guidance for the future.

SOME GENERAL FACTORS IN INNOVATION SUCCESS

Reports resulting from international conferences held in the late 1960s and during the 1970s by the United Nations Office of Economic and Commercial Development summarized the factors believed to be pertinent to innovation successes in the United States. These include:

1. the presence of technologically oriented universities geographically located so that a business climate encourages the cooperative generation of new ventures;
2. entrepreneurs who have previously successful entrepreneurs as examples to follow;
3. the existence of institutions and venture capital sources comfortable with technologically oriented innovators and possessing the rare business appraisal capabilities needed to translate inventions into profits; and
4. good communication networks provided by close proximity to and frequent consultation among all essential personnel in the innovation process.

EXAMPLES OF SUCCESSFUL INNOVATIONS

So many factors are involved that pinpointing any single one or even several major ones responsible for a particular successful innovation is next to impossible. Similarly, it would be foolish to follow slavishly in the footsteps of a successful venture since the rules of the game change, and the players and markets are different the second time around. However, a few examples of successful innovations may illustrate some of the reasons they prevailed.

3M Company—An article in *Innovation* (September 1969) states that 3M is considered one of the best-managed industrial companies in the United States, compiling a remarkable growth over the thirty-year period from 1940 to 1970. Sales increased 120 times, earnings increased at an annual rate of over 13 percent, and market value rose at an annual rate of 18 percent. These results reflected 3M's market philosophy of "look for the uninhabited markets." By 1969 approximately 25 percent of its sales were from products developed in the previous five years. In large part its success came from an exceptional ability to find and develop entrepreneurs from within the company. Market analysis at 3M goes hand in hand with the evolution of product ideas. Entrepreneurs from 3M evolve to be, perhaps, 25 percent technical expert and 75 percent entrepreneur.

Masers and Lasers—Working on a grant from the Department of Defense, Charles H. Townes in 1951 conceived of a means for amplifying electromagnetic radiation that produces coherent beams of microwaves and light—now known as masers and lasers. A patent covering the initial invention was issued in 1959, to be followed over the next two decades by a large number of succeeding patents to cover modifications and improvements. Over 100 companies became involved in developing the technology, which, at first, was devoted to military uses. Nonmilitary products did not appear for some ten years after the initial patent was published. Today, the many uses of these devices have become a multibillion-dollar industry whose greatest impact caused a revolution in both land based and satellite communications—to say nothing of check-out counters in supermarkets!

Platinum Based Antitumor Drugs—Supported by both federal and industrial grants, Barnett Rosenberg in the late 1960s discovered that certain chemical complexes of platinum suppressed reproduction, but not growth, of mammalian tumor cells. Patents were obtained covering these materials and their use, and they were subjected to extensive toxicity, teratogenetic, and clinical testing under both government and industrial auspices. A way was found to overcome their initial high toxicity, and they were finally brought to market late in 1978 by Bristol Laboratories after the long and expensive development period required to obtain Food and Drug Administration clearance. Today these products have a commanding position in the worldwide treatment of many intractable cancers. In addition, great scientific and industrial interest has been spawned in the search for therapeutically active analogs based on platinum or other precious metals.

Mushroom Nutrient—As a result of an intensive study of the nutritional requirements of mushrooms, L. C. Schisler developed a feeding formulation and procedure for its use which greatly improved the quality and yield of commercially produced mushrooms. The patent covering this invention was licensed to a partnership that formed a new venture to manufacture and market the nutrient formulation. The product enjoyed almost immediate acceptance and is currently used extensively by the mushroom industry in the United States and Canada, and good prospects for foreign sales are also evident. The venture is now developing additional related products and services that are expected to form a solid basis for future growth and long-term viability.

EXAMPLES OF INNOVATION FAILURES

There are uncountable reasons why promising ideas never reach the marketplace or are withdrawn after initial market penetration. It would be fruitless to make a comprehensive listing, but a few examples of failed innovations may be illustrative and illuminating.

The New World Computer Company—This company went public in 1978 and, additionally, raised $3.4 million. It is now short of cash for a reason not uncommon among pioneering technology companies—constant dissatisfaction with its products. After developing an excellent computer drive, the cofounding partners decided the drive had insufficient capacity; so, instead of going into manufacturing, they went back to the development laboratory; later they decided to miniaturize their product; next, they entered into an unfruitful joint venture with an overseas company; and now they have acquired another company and its entrepreneurial president. While there is still hope for the future, the company currently has no products on the market, and its credibility has been seriously damaged.

Prestressed Concrete for Highways—In July of 1983, the *New York Times* carried a news item describing the possible use of prestressed concrete for highway construction. The procedure uses about 33 percent less cement, 50 percent less steel, and results in lowered maintenance, fewer cracks, and fewer potholes compared with current methods for the construction of concrete highways. The Federal Highway Administration is preparing a design and construction manual which will authorize the concrete's use for highways. Although test results in the United States and Europe have demonstrated its effectiveness, neither the cement nor the steel industries have shown any interest in commercializing the process. Neither industry perceives increased profits nor other benefits to individual companies. Prestressed concrete experts summarize the situation by commenting that new ideas need a group to promote them, and this idea has no such group nor any other driving force behind it. Furthermore, state and federal highway officials, who should be expected to use the idea as a major cost saver, are ultraconservative and unwilling to take the risk of using anything new when they know the old ways so well. This idea obviously needs an entrepreneur with a knack for convincing die-hard suppliers and customers of the substantial societal benefits of the idea, as well as to assemble a new venture that can become profitable marketing this process.

Synthetic Perfume Bases—The inventor of a useful chemical procedure to produce synthetic base materials for the perfumery and flavoring industries formed a new venture to develop and market these chemicals. Lacking management and marketing expertise, he formed a cooperative undertaking with an experienced entrepreneur, purchased production facilities, and formed a relationship with an experienced marketing organization. These moves overextended his financial capabilities, and a market for his products could not be developed quickly enough to produce an adequate cash flow that could sustain production. As a result the company was forced into bankruptcy.

Ion-Exchange Strengthening of Glass—This process was envisioned as a means for strengthening glass for automobile windshields and architectural use. Although extensive development was performed, the expense of the final product was not competitive with the plastic-laminated glass commonly used for these same purposes. A ruling by the Food and Drug Administration in the early 1970s that mandated the use of shatterproof material in eyeglass lenses revived the technology. However, subsequent development of moldable crack-resistant plastic lenses has limited drastically the market for the strengthened glass. Prospects for development of new uses for this process in the future remain dim.

Summary and Conclusions

A firm base of scientific information and a deep understanding of the marketplace are the essential requirements for a successful innovation. Inventors, entrepreneurs, and businesspeople—each with their special talents and expertise—are needed to put the process into motion and to bring it to a successful conclusion. The innovation process is complex and dynamic, full of pitfalls and opportunities, and subject to many influences—internal and external, predictable and unpredictable. To negotiate a successful outcome requires unusual ideas and outstanding people working together with dedication and goodwill for a common benefit.

Niels Reimers

5

The Government-Industry-University Interface:

Improving the Innovative Process

Introduction

As a society moves from an agrarian to an industrial to an informational economy, the interest in innovation quickens. Specific factors that increase our concern about innovation abound in the United States. Basic major industries, such as steel and automobiles, falter, and increasing quantities of foreign goods appear in our marketplace. We observe a year-by-year continuation of a negative balance of payments and high unemployment.

This concern about innovation is evident in other industrialized countries as well. While Japan might be considered to be an exception to this worldwide concern, it is redoubling its efforts to encourage innovation. In addition, Third World countries are making efforts to enhance their ability to compete in world trade through innovation. Their comparatively low wage scales become competitively advantageous for fewer products

---

NIELS REIMERS *is the founder and director of the Office of Technology Licensing at Stanford University. Mr. Reimers is the past president of the Licensing Executive Society, U.S.A. and Canada. Previously, he was with Ampex Corporation and Philco-Ford Corporation. Mr. Reimers has lectured and written numerous papers on the licensing of basic technology.*

as automated production reduces the labor component of the end product.

Even those countries which, because of their high level of natural resources in comparison to population (such as the U.S.) have been able to enjoy a comfortable standard of living must also look toward innovation as a means to compete in future world trade as their resource-to-population ratio diminishes.

Countries are seeking to leverage their intellectual capacity through innovation and production of high-technology products. In this search for innovation, the linkage between the basic research discovery and the commercial product or process is of particular interest. This chapter intends to review some factors that affect innovation in the United States at the three-way interface between the university, industry, and government.

Overview

There are changes to be made to improve U.S. innovation at the interfaces between government, industry, and universities, even though, on the whole, the system works well enough to be the envy of other countries which seek to find their own formulas for innovation. The government's primary contributions to U.S. innovation are indirect, such as tax policy and support of research programs at American universities. The government's direct involvements, for example, the synfuel program, are, by and large, unsuccessful and divert resources better used elsewhere.

Spin-off innovation from present military and space programs, as well as national laboratories, appears modest; justification for these programs must be based on rationale other than contribution to U.S. commercial competitiveness. Diversion of national, human, and financial resources to the world's largest military program appears the greatest governmental influence on U.S. commerce. That the growing concern for military security may have a more subtle effect, the eroding of our national optimism, and hence innovative spirit, is a thesis briefly explored.

Technology and information controls in the United States seem to be increasing. Much debate ensues as to whether such controls, often confusing and subject to frequent change, are or are not helpful to national security. There is little debate on another effect of rigid controls; scientific research, the underpinning of innovation, does not flourish under secrecy.

A number of university-industry linkages are reviewed. Much public attention has been given to initiatives such as university research parks, patent licensing, and research collaborations with industry. But the greatest contributions to U.S. innovative capacity come in two forms. First, and most significant, is the university graduate who brings to industry the fruits of training in university research programs, overwhelmingly funded by federal and state governments. The research findings of the faculty and students are the second major contribution of universities to U.S. innovation. These findings are promptly and openly disseminated through various forms such as journals, conferences, seminars, industry affiliate programs, and, yes, through the graduated student.

Industry, the central participant in the process of innovation, delivers the end result, making use of the welcome resources (such as the graduated student) that society provides and overcoming the unwelcome impediments (such as technology export controls) that society imposes. On balance, those resources have been positive; however, though U.S. industry has led the world in innovation, signs suggest that this lead is slipping.

Overspecification, overregulation, and rigid planning produce corporate environments not helpful to innovation. Innovation appears to flourish more in less-structured ("skunkwork") operations, as will be discussed later.

The Historical Role of the U.S. Government in Innovation

The Magna Carta of innovation was certainly the Statute of Monopolies passed in 1624 by the English Parliament. This law prohibited all monopolies and restraints of trade. But it recognized patent monopolies were important both to reward the inventors and to promote technical progress and innovation in society. Other countries desiring to enhance economic freedom and growth adopted similar statutes. Except for this legislation, governments in most free-market economies appear to have had little direct influence on innovation and industrial growth.

As America entered the twentieth century, "trust-busting" anti-trust legislation was enacted to curb monopolies which constrained competition, controlled prices, and had a deadening effect on innovation. Otherwise, the government was relatively quiescent with respect to innovation until the Great Depression of the 1930s, when regulatory creep began. This changed dramatically with the advent of World War II when the federal government realized the value of research and development, and corporate contracts and basic research grants in universities became dominant support factors.

GOVERNMENT SUPPORT OF RESEARCH

Following the war, the Office of Naval Research developed an efficient and effective program of grants to university scientists. This program, extended by other government agencies, produced the scientific and technical manpower necessary for "high-technology" industries, as well as many of the scientific discoveries that underlay the important innovations made in mid-century.

Today government is such a pervasive factor in research and development that its *overdirected* involvement could harm innovation. Countries with planned economies (such as the Soviet Union) have lagged far behind market oriented economies in commercial innovation. With few exceptions (such as the space program), the attempts that the U.S. government has made in "directed" innovation have been notably unsuccessful also (e.g., the synfuels program). Frederick-Carl Beier, director of West Germany's Max Planck Institute for Foreign and International Patent, Copyright, and Competition Law, has suggested that a motto for the appropriate balance should be: "only as much government as absolutely necessary and as much private industry as possible."

Current worldwide competition requires continued government involvement in support of university research to produce scientific and technical manpower, as well as to enable the breakthrough innovations resulting from the research that allow the U.S. to compete successfully in world trade. Government can assist innovation in a free-market system through the direct support of research, support of graduate education, and indirect market incentives. An example is the establishment of needs that results in a market "pull" toward technological solutions, rather than a government planned "push."

Major innovations or "breakthroughs" often arise from undirected research. But, as Ralph Gomory, vice president and director of research at IBM Corporation, noted (in a May 6, 1983, article in *Science* magazine): "Real breakthroughs do occur; they are rare and stunning events. The more common course of technological evolution is steady year-to-year improvement, and when that is rapid and persistent, the results are just as revolutionary."

A Path of Innovation

As long ago as 1968, the National Science Foundation sponsored a systematic study of the role of research findings in the overall process leading eventually to a major technological innovation. Titled *Technology in Retrospect and Critical Events in Science* (TRACES), it was prepared by the Illinois Institute of Technology Research Institute and later extended by Battelle-Columbus Laboratories. This study retrospectively examined the key technological events which led toward major innovations. In the TRACES cases, the average time from conception to demonstration of an innovation was nine years. Of the key events, approximately 70 percent were nonmission research, 20 percent mission oriented research, and 10 percent development and application.

Nonmission events along the path to the electron microscope were discoveries by Maxwell, Planck, Roentgen, Hertz, and others (see Figure 1). Ultimately, in 1937, Metropolitan-Vickers produced the first commercial electron microscope. This was followed in 1939 by the first commercial unit to exceed the light microscope capabilities, manufactured by Siemens. In 1940, RCA made the first commercial unit in the U.S.

While the key events largely took place in universities and industrial laboratories, government support of universities, as well as tax and other policies, contributed to the innovation and commercialization of the electron microscope.

Recent experience in U.S. universities indicates that the majority of the technological developments now occurring have a

Fig. 1. The Electron Microscope

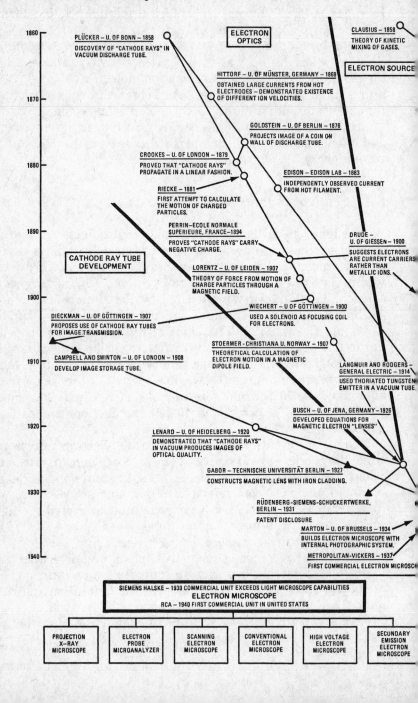

ELECTRON OPTICS

ELECTRON SOURCE

1860 —
PLÜCKER – U. OF BONN – 1858
DISCOVERY OF "CATHODE RAYS" IN VACUUM DISCHARGE TUBE.

CLAUSIUS – 1858
THEORY OF KINETIC MIXING OF GASES.

HITTORF – U. OF MÜNSTER, GERMANY – 1869
OBTAINED LARGE CURRENTS FROM HOT ELECTRODES – DEMONSTRATED EXISTENCE OF DIFFERENT ION VELOCITIES.

1870 —

GOLDSTEIN – U. OF BERLIN – 1876
PROJECTS IMAGE OF A COIN ON WALL OF DISCHARGE TUBE.

CROOKES – U. OF LONDON – 1879
PROVED THAT "CATHODE RAYS" PROPAGATE IN A LINEAR FASHION.

1880 —

EDISON – EDISON LAB – 1883
INDEPENDENTLY OBSERVED CURRENT FROM HOT FILAMENT.

RIECKE – 1881
FIRST ATTEMPT TO CALCULATE THE MOTION OF CHARGED PARTICLES.

PERRIN–ECOLE NORMALE SUPERIEURE, FRANCE–1894
PROVES "CATHODE RAYS" CARRY NEGATIVE CHARGE.

1890 —

DRUDE – U. OF GIESSEN – 1900
SUGGESTS ELECTRONS ARE CURRENT CARRIERS RATHER THAN METALLIC IONS.

CATHODE RAY TUBE DEVELOPMENT

LORENTZ – U. OF LEIDEN – 1907
THEORY OF FORCE FROM MOTION OF CHARGE PARTICLES THROUGH A MAGNETIC FIELD.

1900 —

WIECHERT – U OF GÖTTINGEN – 1900
USED A SOLENOID AS FOCUSING COIL FOR ELECTRONS.

DIECKMAN – U. OF GÖTTINGEN – 1907
PROPOSES USE OF CATHODE RAY TUBES FOR IMAGE TRANSMISSION.

STOERMER - CHRISTIANA U. NORWAY – 1907
THEORETICAL CALCULATION OF ELECTRON MOTION IN A MAGNETIC DIPOLE FIELD.

1910 —
CAMPBELL AND SWINTON – U. OF LONDON – 1908
DEVELOP IMAGE STORAGE TUBE.

LANGMUIR AND RODGERS – GENERAL ELECTRIC – 1914
USED THORIATED TUNGSTEN EMITTER IN A VACUUM TUBE.

BUSCH – U. OF JENA, GERMANY – 1926
DEVELOPED EQUATIONS FOR MAGNETIC ELECTRON "LENSES"

1920 —
LENARD – U. OF HEIDELBERG – 1920
DEMONSTRATED THAT "CATHODE RAYS" IN VACUUM PRODUCES IMAGES OF OPTICAL QUALITY.

GABOR – TECHNISCHE UNIVERSITÄT BERLIN – 1927
CONSTRUCTS MAGNETIC LENS WITH IRON CLADDING.

1930 —
RÜDENBERG-SIEMENS-SCHUCKERTWERKE, BERLIN – 1931
PATENT DISCLOSURE

MARTON – U. OF BRUSSELS – 1934
BUILDS ELECTRON MICROSCOPE WITH INTERNAL PHOTOGRAPHIC SYSTEM.

METROPOLITAN-VICKERS – 1937
FIRST COMMERCIAL ELECTRON MICROSC[...]

1940 —

SIEMENS HALSKE – 1939 COMMERCIAL UNIT EXCEEDS LIGHT MICROSCOPE CAPABILITIES
ELECTRON MICROSCOPE
RCA – 1940 FIRST COMMERCIAL UNIT IN UNITED STATES

| PROJECTION X-RAY MICROSCOPE | ELECTRON PROBE MICROANALYZER | SCANNING ELECTRON MICROSCOPE | CONVENTIONAL ELECTRON MICROSCOPE | HIGH VOLTAGE ELECTRON MICROSCOPE | SECONDARY EMISSION ELECTRON MICROSCOPE |

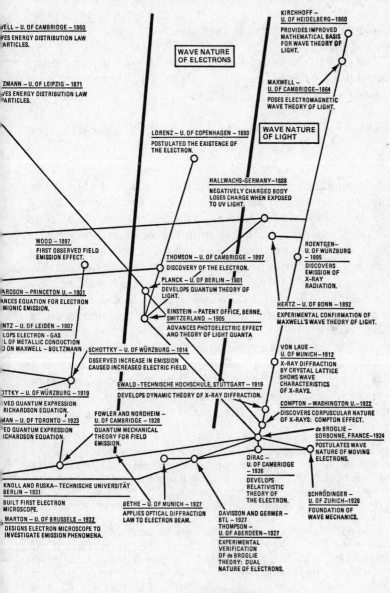

significantly shorter transition time from conception to commercial use. However, *major* breakthroughs continue to experience a seven- to ten-year gestation from discovery to significant commercial application.

The TRACES study provides evidence of the continuing significant interplay of basic and applied research, ultimately realized in commercial products.

Innovation from U.S. Military and Space Programs

VIEW FROM JAPAN

Benefits from the spin-off of innovations from military and space programs are frequently asserted. However, Masanori Moritani, at the Japanese Nomura Research Institute, was unable to confirm the existence of a significant number of such innovations resulting from these programs, except for those occurring during a short period at the initiation of the U.S. space program. In his book, *Japanese Technology,* Moritani claims that concerns by Japanese businessmen that technological spin-offs from the U.S. military and space programs make U.S. private industry a more formidable competitor are "unwarranted." He argues instead that the military and space programs were a detriment to U.S. competitiveness by siphoning off most talented researchers and engineers from corporate enterprise. "The stagnant steel, consumer electronics, and general machinery industries are unable to compete for top-caliber researchers," Moritani declares. Continuing, he states:

> In Japan, in sharp contrast, the top Japanese researchers and engineers are committed to the development of civilian technology, not just in the computer field, but in VTRs [video tape recorders] and VTR cameras, televisions, automobiles, steel, and the like. The competence of researchers at the top does not differ that much from country to country. Even China has developed a hydrogen bomb despite the backwardness of its industrial technology, and has launched as many as eight satellites. The difference lies in how this top class is put to use. Perhaps the greatest benefit Japan has gained from taking shelter under America's nuclear umbrella for a "free ride" in defense has not been financial but human. Thanks to America, Japan has not had to squander its most talented engineers in the development of military technology.

Military Demands May Erode Ability to Innovate—America's growing concern for military security (beginning in the late 1940s) may well be debilitating to the basic optimism necessary for innovation. Professor David Kennedy, in a *Stanford Magazine* article, "War and the American Character," suggests that this continuing intrusion of military preparations into American life accounts for the steady erosion of our exuberant optimism as a people. By historic and international standards, the U.S. flourished for 175 years in a setting relatively free of war and military preparation. Kennedy perceives that, while history is catching up with us, we are still distinctive among most nations in never having experienced the terror, demoralization, and destruction of modern war in our heartland.

> In this important aspect, we are still innocent in a way that sets us apart from nearly all peoples in Europe or Asia. It is not pleasant to ask what would happen if America one day became the battleground. . . . Given the long lines of internal transportation and communication, the concentration of our population in vulnerable metropolitan areas, the generally comfortable lives to which so many have become accustomed, the racial and economic conflicts that only relative peace and prosperity have made manageable, and the deep American hostility to authority and coercion that necessary martial law would entail, it is especially frightening to contemplate the circumstances in which America would lose this last item of innocence about modern warfare.

Kennedy further suggests that the fact that the U.S. has not felt the "direct pain of war" in its heartland may explain the "long-deep acquiescence" of the country to the war in Southeast Asia.

Even our most deadly conflict, the Civil War, reinforced popular attitudes in the North that war was waged in remote areas, accelerated economic growth, and strengthened social elites. The North's victory confirmed an attitude of "righteousness and omnipotence" toward others, providing "a huge repository of self-congratulation on which the nation has drawn ever since." These attitudes, continues Kennedy, were reflected in Woodrow Wilson's call for the U.S. to make the world "safe for democracy" in World War I and in Franklin Roosevelt's demand for unconditional surrender in World War II.

These global conflicts erased a long-held, romantic notion of battle for millions of Americans. But our casualties were only 1 percent of Europe's, our industrial output soared, and wealth poured into America "on a scale that invited comparison with the oil exporting nations today," Kennedy says. But the sense of buoyance and optimism that characterized earlier epochs was noticeably missing. "Books like *The Naked and the Dead, The Caine Mutiny,* and *From Here to Eternity* were almost shot through with a sense of futility, absurdity, and resignation." These views reflect "deep anxieties about security in an age in which the U.S. has suddenly become intricately involved in a volatile, unstable world order—an experience for which 175 years of 'free security' left the American psyche peculiarly unprepared."

"So, too, with the economic abundance and economic status of America," he continues. Military expenditures averaged less than 1 percent of our gross national product well into the start of the twentieth century, but

> . . . since 1950 [that fraction has increased, and] we now spend more than any other nation on military items. . . . [Vietnam alone cost $330 to $350 billion.] We now know the constraints on individual freedom, the dark uncertainties of the spirit, the poisonous effect on our political life that war and preparation for war has long made familiar in other societies, but from which we were for so long spared.

This relentless pressure on the American psyche stemming from fears necessary to accept preparation for war, the possibility of a nuclear Armageddon, and the very presence of war erodes the national self-confidence. If Kennedy's assertion is correct (that there is an erosion of national self-confidence) and if we accept that creativity and innovation are characteristic of a confident and optimistic people, we should be concerned that our ability to innovate may also be quietly eroding.

National Security and Export Controls

The national security restrictions on transfers of technology and information are many and varied. There are several major laws under which our government may act to restrict technology exports and many other influences which affect implementation of those laws and future technology controls. These controls in general do not enhance innovative activity.

Atomic Energy Act—One of the oldest statutory authorities is the Atomic Energy Act (42 U.S.C. 2011–2296). Originally adopted in 1946 and significantly overhauled in 1954, this law controls exports of restricted data concerning the design, manufacture, or utilization of atomic weapons; the production of special nuclear material; or the use of special nuclear material in the production of energy. The law has been invoked against private parties who independently developed information of this type.

Another law restricting technological information was enacted in 1981 as an amendment to the Atomic Energy Act (42 U.S.C. 2168). It permits the secretary of energy, in certain cases, to prohibit the dissemination of unclassified information pertaining to the design of production facilities, security measures for the protection of production facilities and nuclear material, and the design of any atomic weapon or component even if such information has been declassified.

Gerald Lieberman, Vice Provost at Stanford University, commented to the Department of Energy (DOE) in April 1983 that, contrary to the intent of Congress, proposed DOE rules on unclassified nuclear information have "unlimited potential to chill research, teaching, and the general interchange of information." He observes the traditional freedom to publish the results of university research and, further, that the Atomic Energy Act of 1952 provides "the dissemination of scientific and technical information relating to atomic energy should be permitted and encouraged." The proposed regulation "is so vague, ambiguous, inconsistent, and couched in such general categorical terms as to be capable of interpretation [which would give] the Secretary [of Energy] maximum flexibility to prohibit dissemination of anything and everything he chooses. . . ."

Invention Secrecy Act—Another statute which the government may use to control information developed independently by the private sector is the Invention Secrecy Act of 1951 (35 U.S.C. 181–188). The law dates back to America's entry into World War I, but the present statute was enacted in 1952. This act provides that the patent commissioner shall make a patent application available to U.S. defense agencies for review whenever, in his opinion, the publication or disclosure of an invention might be detrimental to the national security, even if the government does not have a property interest in the patent. If a defense agency determines that the publication or disclosure of the invention would be detrimental to national security, the commissioner shall order that the invention be kept secret and shall withhold the grant of the patent for not more than one year, subject to renewal of his order.

Arms Export Control and Export Administration Acts—Two sets of major laws control the export of technical information. One of these is the Arms Export Control Act (22 U.S.C. 2751–2794), and the other is the Export Administration Act (50 U.S.C. App. 2401–2420). These laws not only govern the export of data and goods from the U.S., but they also limit the access of foreign nationals to such information and materials within the United States. Agency regulations implementing these statutes embrace scientific information and define exports broadly enough to include the domestic publication or release of information.

The Arms Export Control Act governs the sale of U.S. defense articles, services, and technology abroad. The Export Administration Act controls the export of goods and technology which would make a significant contribution to the military potential of any country or combination of countries and which could prove detrimental to national security of the United States. The Department of Commerce administers the Export Administration Regulations and the Department of State administers the International Traffic in Arms Regulations under these laws. These departments and the Department of Defense consult each other on sensitive license applications under either set of regulations, but while the Department of Commerce has expediting procedures, the Department of State does not. The Department of Defense uses its "Military Critical Technologies List" as a reference for making recommendations to either the Department of Commerce or State. The 1983 list covers about 700 pages, and part of it is classified.

More than 90 percent of U.S. exports, in terms of dollar value, are shipped under general license authorization without the need to obtain a validated export license in advance. However, most experienced exporters know which items raise national security concerns, and they do not attempt to procure validated export licenses for them.

International Exchange and Proscriptions—The Coordinating Committee on Export Controls (CoCom) consists of fourteen of the fifteen countries in the North Atlantic Treaty Organization (Iceland is not a member), plus Japan. The body has no official power to prohibit sales by its member countries, but its recommendations are often, though not always, followed. The United States controls some items that other CoCom members do not.

CoCom has routinely granted approximately 1,700 exceptions to its rules each year.

Congressional Jurisdiction—In the Senate alone, numerous committees have overlapping jurisdictions on technology transfer. The Senate Committee on Banking, Housing, and Urban Affairs has jurisdiction over criminal sanctions to enforce export controls. The Committee on Foreign Relations has jurisdiction of the Arms Control Act. The Committee on Governmental Affairs reviews the ability of the executive branch to enforce export controls. The Senate Committee on Intelligence is often consulted in the preparation of hearings by other committees on these subjects and the Committee on Armed Services has an obvious interest in Department of Defense matters.

Freedom of Information Act—Further legislation upon the flow of technological information is the sunshine laws, whose intended purpose is to open government files to public scrutiny. The Freedom of Information Act (FOIA) provides a number of exceptions for certain classes of information that need not be released under an FOIA request. For example, in some cases, release of information may be delayed in order to allow patents to be filed when premature public release would bar patenting.

Legislation to amend the FOIA has been proposed that will allow agencies to withhold "technical data" that may not be exported lawfully outside of the U.S. until the appropriate approval or license has been granted. Other proposed legislation would deny foreign entities the right to obtain, under an FOIA request, documents from government agencies.

Restrictions on Information Flow: "The CIA Report"—Much debate over the transfer of technology surrounds a published report of the U.S. Central Intelligence Agency (CIA) entitled *Soviet Acquisition of Western Technology* (1982). The thrust of the report is that the success of Soviet and East European intelligence services in acquiring U.S. technology has resulted in a significant threat to American security. Although the report states that the "overwhelming majority" of militarily significant technology was acquired by intelligence organizations, the CIA believes open and legal acquisitions are still important because "it is often the combination of legally and illegally acquired technologies that gives the Soviets the complete military or industrial capability they need."

One example the CIA gives of legal channels used by the Soviets is the Dressler project, which supplied three foundries for the Soviets' Kama River truck plant. The CIA asserts that large numbers of trucks produced there in 1981 are now being used by Soviet forces in Afghanistan and by Soviet military units in Eastern Europe opposite NATO forces. Another example of legal acquisitions discussed is the Soviet purchase of 168 grinding machines for the production of small, high-precision bearings. The CIA claims these purchases provided the Soviets with the capability to manufacture precision bearings in large volume sooner than they could have on their own. Defense officials argue this sale enabled the Soviets to speed construction of more stable and accurate missiles having a multiple-warhead capability.

A third example is the legal acquisition by the Soviet Union of two huge, floating dry docks purchased from the West for civilian use and diverted to military purposes. When the Soviets took possession of one of the dry docks in 1978, they used it for their Pacific Naval Fleet. The other was sent to the Northern Fleet in 1981. According to the CIA, these are the only dry dock facilities in either of the two major Soviet fleet areas, northern or Pacific, capable of servicing the new Kiev-class aircraft carriers. Their importance will be greater when the Soviets construct the still larger carriers for high-performance aircraft projected for the 1990s.

National Academy of Sciences Report—On September 30, 1982, a special panel of the National Academy of Sciences issued its own findings on the transfer of U.S. technology. The Panel on Scientific Communication and National Security concludes: "There has been a substantial transfer of U.S. technology—much of it directly relevant to military systems—to the Soviet Union from diverse sources." However, "there is a strong consensus . . . that universities and open scientific communication have been the source of very little of this technology-transfer problem." The panel emphasizes that national security is more apt to be enhanced by a policy of open scientific exchange that promotes scientific accomplishment than by a policy of secrecy controls.

Proponents of selective national security restrictions on technological information offer a counter to the National Academy of Sciences report. They argue that advances in technological innovation and economic productivity occurred during the very years in which rather strict controls have been in effect. They further claim that many of the most successful and innovative corporations are those that deal extensively in areas of national security information restrictions and themselves engage in additional industrial security practices. They maintain that there is little persuasive evidence of economic damage or innovation chill due to selective applications of national security controls.

"Secrecy: The Road to Nowhere"—Edward Teller, who played a major role in development of the hydrogen bomb, claimed in M.I.T.'s *Technology Review:*

> In the last third of the century, the United States has lost its position in all military fields, most specifically in those where we practice secrecy. . . . We now have millions of classified technical documents; we also have falling productivity. Rapid progress cannot be reconciled with central control and secrecy. The limitations we impose on ourselves by restricting information are far greater than any advantage others could gain by copying our ideas.

Teller, a consultant to Lawrence Livermore Laboratory, has fought to open up the classification system for government research laboratories. He points out that technical fields where the U.S. leads, such as electronics, are those "where we practice the most openness."

The San Diego Incident—In August of 1982, four Soviet scientists were to attend the annual meeting of the Society of Photo-Optical Instrumentation Engineers in San Diego. Their attendance triggered actions by the Departments of Commerce, State, and Defense and ultimately led to withdrawal of some 150 papers by U.S. scientists from the meeting. Penalties for an individual's knowingly violating technology export laws include up to ten years in prison and fines up to $250,000.

Cryptography and the NSA—A voluntary and self-policing system for university researchers in cryptology evolved in the late 1970s as a result of concern by the NSA and CIA of sensitive cryptographic technology being transferred to the Soviet Union. While recognizing the importance of unfettered scientific communication, Admiral Bobby Inman, then speaking for the CIA, expressed his belief that the problem of leakage from academics, while then small in comparison to industry and espionage related leakage, would be a growing problem.

University–DOD Forum—Five university presidents (Stanford, Cal Tech, Cornell, M.I.T., and UC—Berkeley), concerned about the evolving constraints upon international scientific communication, wrote a letter which led to establishment of a joint university —DOD forum to study the question of science and technology transfers at universities. Their letter stated, "Restricting the free flow of information among scientists and engineers would alter fundamentally the system that produced the scientific and technological lead that the government is now trying to protect and leave us with nothing to protect in the very near future." The letter goes on to say that the significant practical difficulties of enforcing export restrictions are "virtually impossible" for universities to administer. It is difficult to imagine guards at classroom doors of U.S. universities, enforcing security by checking students who would be required to wear badges indicating their clearance to attend certain lectures.

Technology Hemorrhage—Concerned officials have described Soviet access to sophisticated U.S. devices as a "hemorrhage of technology." Democratic Senator Sam Nunn of Georgia suggests that the United States is funding two military research programs —our own and the Soviets'. Democratic Senator Paul Tsongas of Massachusetts considers certain technology controls absurd. "We lose the technology, the foreign business, and become known as an unreliable supplier," he notes. Boeing was denied approval to sell to Ethiopia a 767 aircraft with a laser gyro. Ethiopia then bought a French Airbus with the same laser gyro, which could be sold to France, an ally, by the American manufacturer.

President's Office of Science and Technology Policy (OSTP) Report—A government-wide study group headed by OSTP is expected to soon release its report of a study of national security and technology transfer issues. The study focuses on government organizational structure concerned with technology transfer matters, U.S. policies on controls or lack of controls or technology transfers to various nations, and issues associated with unclassified but militarily sensitive data. Louis T. Montulli of OSTP, in describing the government's view of the issues, has reported that "right now, forty-four separate groups in ten or more U.S. departments are either studying this subject or actually executing the present policy."

Federal Acquisition Regulations (FAR)—The FAR are new, uniform regulations to be used by government agencies for procurement. The final section, covering copyrights and technical data, was offered for public comment in May 1983. According to one of the reviewers, "Not only are there unacceptable controls of freedom of publication, inappropriate 'backdoor' enforcement of export controls, but, through the copyright and data clauses, the tenets of PL 96–517 [the University and Small Business Innovation Act] are violated."

New Technology Control Laws—Legislation to replace the Export Administration Act of 1979 was introduced in the Ninety-eighth Congress. In S. 434 (the Garn Bill) "technology" is defined broadly enough to include virtually any information or goods as being subject to government control. Senate Bill 407, introduced by Senator Nunn, gives criminal enforcement power to the Customs Service as well as statutory authority for warrantless arrest and search and seizure. A second bill by Senator Nunn, S. 408, entitled the Technology Securities Enforcement Act of 1983, stretches racketeering laws to cover violations of the Export Administration Act and Arms Export Control Act, exposing violators to a twenty-year prison term. S. 408 also amends electronic surveillance statutes to permit court-authorized surveillance where there is probable cause to believe that a violation of the Export Administration Act, the Arms Export Control Act, or the new technology theft statute is being committed.

INNOVATION, SECRECY, NATIONAL SECURITY,
AND TECHNOLOGY CONTROLS

One can debate whether or not broad controls of technology and information enhance American security. There is little debate, however, that scientific research and innovation do not flourish under secrecy. Recall that the patent system in England was established by Parliament in 1624 because the practice of secrecy was inhibiting technical progress and innovation. Further, Article I, Section 8 of the U.S. Constitution provides for a patent system for the same reasons.

Regulations, Specifications, and Special Interests

Regulation vs. Innovation—"Regulation creep" is a disease that can have deleterious effects on innovation. It is progressive in nature, appearing to attack older societies more severely than younger ones. Rigid regulations and specifications for government and industry procurement can serve to dampen innovation and increase costs. Creativity is unlikely to flourish in such an environment. However, there may be a limited number of situations where regulation can spur innovation. For example, tightening automobile exhaust emission standards acts as a regulatory "pull" for new and improved methods and devices for lowering exhaust emissions.

Skunkwork vs. Specification—Thomas Peters, of the Stanford Graduate School of Business, has reported numerous anecdotal cases to justify his assertion that small skunkwork operations in a company will time and again provide more successful results than project teams operating under detailed specifications. While noting the word *skunkwork* may have originated with L'il Abner, Peters believes *skunkwork* apparently was used first as a business term to identify a group of Lockheed mavericks who produced the U-2 aircraft. "When a practical innovation occurs, a skunkwork, usually with a nucleus of six to twenty-five, was at the heart of it. The skunkwork seldom reinvents the wheel," claims Peters. Some general sense of direction may help, such as "Northwest." "What's not sensible," he argues, "is trying to prespecify the difference between a course of 343 degrees and a course of 346 degrees."

In their bestseller, *In Search of Excellence: Lessons from America's Best Run Companies,* Peters and Robert Waterman point out that the "best run" companies have provided the environments that stimulate skunkwork teams. Even giant IBM turned to a collection of no less than seven parallel skunkwork teams to develop its enormously successful computer.

Strategic and Product Planning—James Utterback at M.I.T., from his studies of many successful products, concludes that "the initial use and vision for a new product is virtually never the one that is of the greatest of importance commercially." There is an apparently inherent organizational tendency to do the wrong thing vis-à-vis stimulating innovation. "In 32 of 34 companies, the current product leaders reduced the investment in the new technology in order to pour even more money into buffering the old," he observes. Neither Utterback nor Peters suggests doing away with strategic or other technology planning. But in Peters's words, a company desiring to encourage innovation needs to allow "maximum play" to the "substantially sloppy process" that produces successful innovations.

Special Interest Coalitions—It is not only governments that are responsible for rules which act to constrain innovation. Mancur Olson, University of Maryland economist, in *The Rise and Decline of Nations: Economic Growth, Stagflation, and Social Rigidities,* theorizes that the special interest coalitions endemic to free societies become more and more influential as a stable and affluent democracy matures, giving rise to a form of national "economic sclerosis." As analyzed by Eliot Marshal, in his 1983 review in *Science,* Olson's theory works as follows:

> In societies that permit free trade and free organization, coalitions will form around marketable goods and services. Groups of producers, like those who grow wheat or own oil, will organize to protect their assets and, if possible, will boost profits by raising prices. Physicians and lawyers do much the same in joining professional societies. Labor unions organize workers to bargain for wages.
>
> In the early stages of this coalition building process, there are relatively few interest groups, and their memberships are small compared to the society in which they operate. As they develop, they try to impose a variety of specialized rules on the economy that supports them. By law or collusive contracts, they make penalties for those who would market the same goods or services outside the group. They also offer selective advantages to those who join and cooperate. Because these groups are small (Olson says they typically include no more than one percent of the people in their state), they have no incentive to boost members' welfare by boosting the state's welfare. Instead, they concentrate on promoting their own narrow interest, even at the cost of retarding the general economy. A modest effort at self-aggrandizement may bring great rewards.
>
> As time goes by, tariffs, price supports, monopoly prices, wage guarantees, and business codes grow more numerous. All are intended to channel commerce for them. *The combined effect is to create obstacles to trade and to prevent innovation.* The economy suffers.

In the past, nations suffering from this affliction have enjoyed renewed growth after a cataclysm has intervened to wipe out existing trade barriers, or when new territory has been opened for development. Sometimes, the power of a domestic group is undercut by low-cost imports, if the imports are not blocked. Rarely has any nation abolished special interest codes voluntarily.

Government Patent Ownership

PUBLIC LAW 96–517

Culminating over ten years of effort toward development of policies that would best encourage the exploitation of the fruits of government funded research, the University and Small Business Innovation Act was implemented in Public Law 96–517 (effective July 1, 1981). This gives nonprofit organizations and small business firms first option to acquire title to inventions conceived by them under federal research funding. As was brought out during congressional hearings, when the government took title to inventions from federally funded research, only 3 to 5 percent of the patented inventions would eventually be commercialized. In contrast, when title was in the name of a university, approximately 50 percent of the patented inventions were eventually licensed to industry for commercialization. Close to 30,000 unlicensed patents had been accumulated by the government at the time the bill was passed. PL 96–517 allows a federal agency to exempt university operated laboratories from the law. The Department of Energy, which administers eight such university laboratories, has chosen to exempt them.

Background Policy for PL 96–517—One of the motives behind the legislation which led to the passage of the University and Small Business Innovation Act was to encourage industry involvement in university research. This required the reduction of the prospect for "contamination" of rights to research results in a laboratory which was funded in part by a government agency. A common occurrence in a laboratory with such mixed funding would be attribution of an invention *both* to a sponsoring company *and* a sponsoring government agency. While the company would have rights to an invention through its sponsorship, the government could assert rights in its independent share of the invention and then make rights in that invention available to the company's competitors who had not participated in any of the costs of the research. This "contamination" was removed by PL 96–517, which provided (with certain exceptions) first option to rights in inventions under government supported research to the universities.

Changes Following Enactment of PL 96–517—In order to ascertain the possible effects of PL 96–517 on university-industry interaction, the author conducted a survey of about twenty universities known to interact actively with industry. Sixteen responses were received. All respondents indicated that university-industry interactions were increasing, although not attributable solely to PL 96–517. Data on the number of invention disclosures during 1978–82 showed a steady increase in the annual rate of invention disclosures made; the largest increase was in 1982—up approximately 20 percent from 1981.

Universities were asked about the change in industry support of 1982 compared to 1978. In all cases, the percentage of the total university research budget supported by industry increased significantly, and several predicted that 1983 would show an even larger increase. Even so, the average share of industry research support at universities is well below 6 percent, and even a larger percentage increase will not provide a substantial addition to or substitute for federal funds.

The Federal Laboratories

FEDERAL LABORATORY BUDGETS

Often overlooked in analysis of factors in research, development, and commercialization in the United States are the national laboratories. The 1979 budget of $794 million for the Sandia and Livermore laboratories alone was greater than the combined 1979 research funding of the top six (in terms of funding) U.S. research universities: The Johns Hopkins University, Massachusetts Institute of Technology, Stanford University, University of Washington, the University of California at San Diego, and the University of California at Los Angeles. Moreover, the 1983 budget of those two laboratories was $1,630 million—double that of 1979 and equivalent to the 1983 federal funding of the research programs of not only the above six universities, but also the estimated funding of the next six as well.

Overall, the federal laboratory system incorporates some 755 laboratories and consumes more than 33 percent of the federal research and development budget. It has been charged that the flow of dollars into the laboratories has been at the expense of industry and university research laboratories, which, ironically, have comparatively superior track records of contributions to U.S. innovation.

WHITE HOUSE SCIENCE COUNCIL REPORT ON THE LABORATORIES

Based on a 1983 report of the White House Science Council, the *New York Times* reported, "The federal laboratory system has 'serious deficiencies' that limit the quality of its work and the nation's ability to compete against foreign technological research." Only three laboratories were praised: The Fermi National Laboratory in Illinois, the Stanford Linear Accelerator Center in California, and the China Lake California Naval Ordnance Laboratory.

ENERGY ADVISORY BOARD REPORT ON FEDERAL LABORATORIES

In a late 1982 report, the Energy Advisory Board criticizes the "floundering" system of support, management, and oversight of those federal laboratories administered by the Department of Energy. On the other hand, in a 1982 article in *Chemical and Engineering News,* a Los Alamos laboratory official is quoted as saying:

One of our problems is that there are too many industrial concerns at the federal trough, and they are competing with each other and the labs. And in many projects it isn't clear that what they do is any different than what the national labs do. If we are going to be assessing the role of labs, we ought to be assessing the whole issue of federal funding, rather than industrial relationships. Many contractors are producing useless gold-plated widgets for the Department of Defense or the Department of Energy, and we ought to take a look at who those guys are.

THE FUTURE OF THE LABORATORIES

The Energy Advisory Board and White House Science Council reports do not recommend closing the national laboratories, but rather they note their potential as important centers of research on national problems. David Packard, chairman of the prestigious White House Science Council panel on the federal laboratories, warns, however, that unless corrective action is taken with regard to the laboratories, the nation will face "serious problems" that will threaten its scientific and technological leadership.

Industrial and University Research

BEFORE WORLD WAR II

In the United States, university based research developed toward the end of the nineteenth century, which is about the same time the modern industrial corporation was emerging. Industrial research laboratories became a feature of prominent U.S. corporations after 1910, reaching a peak in the early 1930s. In 1927, it was estimated that total national research and development expenditures were $212 million. Over 90 percent of these funds was estimated to represent work by industrial concerns in their own research laboratories. A 1982 National Science Board (NSB) report on university-industry research relationships considers the importance of these industrial research laboratories to be that of "having created a locale for advanced research and development, and required staffing by scientists and engineers with advanced training and degrees."

In the early part of the century, very wealthy individuals and large, general purpose foundations, such as The Rockefeller Foundation and the Carnegie Institution of Washington, were sources in aiding research in American universities. More important for the support of research in the basic sciences were the smaller, specialized foundations, such as the Dreyfus Foundation, the Petroleum Research Fund, Research Corporation, and the Alfred P. Sloan, Jr., Foundation.

Through the land-grant system, agriculture related research was encouraged by both federal and state governments. U.S. university enrollments doubled every twenty years from 1900 to 1960, providing a steadily growing, well-educated work force for science and engineering teaching and research.

In the mid–1920s, Herbert Hoover, then secretary of commerce, sought to raise $1 million from American industry to support basic research in the nation's universities. He told industry leaders they would lose a form of intellectual capital if they did not make it possible for able researchers in universities to be relieved of some of their teaching obligations and to be equipped to do first-rate scientific research. This effort failed because of corporate reluctance to contribute to openly published research that could give advantage to competitors. The Hoover campaign did, however, create support for the National Research Council and for a program that kept science alive during the Great Depression.

EARLY UNIVERSITY-INDUSTRY COOPERATION

In the period prior to World War II, the NSB report notes several university programs that were distinguished in their vital approach to university-industry cooperation. Particularly noteworthy was the effort led by William M. Walker, Warren K. Lewis, and Arthur D. Little to develop a chemical engineering curriculum at M.I.T. closely suited to the needs of the chemical industry. Considerable financial support was received from companies through Walker's Research Laboratory for Applied Chemistry. Research support was also received for the aeronautics program established at Cal Tech by Theodore Von Karman. This contributed significantly to the growth of the aeronautical industry.

At the University of Illinois, the chemistry and chemical engineering program of Roger Adams made their chemistry and chemical engineering departments into the world's largest producers of doctorates in any discipline. While this program did not include a major component of direct industrial support for academic research, it provided considerable support for student fellowships and encouraged the flow and exchange of people between the university and industries.

POSTWAR ENHANCEMENT OF RESEARCH SUPPORT

World War II brought together unprecedented numbers of industrial, academic, and government scientists and engineers in collaboration on wartime projects. Notable innovations included radar, penicillin, synthetic rubber, and nuclear energy. These collaborations are enthusiastically described in the NSB report:

> The scientists themselves found the process exhilarating and intellectually exciting. This excitement was also communicated to their graduate students, who learned that product-oriented work can give high intellec-

tual stimulation. In addition, the contacts made and the process broadened student perspectives on their work and career options.

After the cessation of hostilities, the Office of Naval Research in particular became an important factor in developing the research base at universities. Its support enabled leading scientists to re-establish and enlarge research programs earlier sacrificed to the war effort. This support also illustrated the value of relationships of industry and university scientists that lead to many consulting arrangements, as well as direct employment of academics in corporate research laboratories.

Perhaps the most productive of any corporate research laboratories, in terms of scientific discoveries, are the Bell Laboratories. For example, their 1947 discovery of the transistor by William Shockley and others led to a new industry. Bell Laboratories encouraged their scientists to spend sabbaticals at universities and, likewise, enabled university scientists to work at Bell. In addition many science professors encouraged their brightest students to work for a few years in Bell Laboratories' well-equipped facilities before seeking an academic appointment.

IMPACT OF FEDERAL FUNDING

A fundamental shift in emphasis for university research arose in the 1950s and 1960s due to the ever-increasing growth in federal funds for academic science from the National Science Foundation, the National Institutes of Health, and other agencies and departments. This decreased the need for industrial support of university research, gradually led to barriers between university and industry, and sparked negative attitudes on both sides. These differences widened during the period of the Vietnam War. Though by no means universal to all campuses or in all companies, this apparent deterioration of university-industry ties was reversed in the 1970s. Efforts of "bridge building" began, and recognition of the value of interaction between universities and industries increased.

The Sequence of Innovation

Stanford President Donald Kennedy, former head of the Food and Drug Administration, observes that there appears to be a fairly standardized historical sequence of innovation following

World War II and the rise of the modern research university.
He explains:

> The first phase is publicly funded and oriented toward the discovery
> and explanation of basic phenomena. It is characterized by loose informal
> organization, very open communication, including quick publication of
> all details of an experiment. Typical institutions where this study of
> phenomena occurs include departments of biology, chemistry, or physics,
> a laboratory in the NIH institute, or a special industrial organization
> like Bell Telephone Laboratories.
>
> The second phase is best called application. It is focused upon processes,
> and takes place in various settings: applied institutes, some university
> departments (of engineering, for example), nonprofits (like SRI Inter-
> national or Battelle), and industrial laboratories. There is a mix of public
> and private funding and environments that are variable with respect to
> proprietary secrecy.
>
> In the third stage, development, attention is given to practical applica-
> tion, including such matters as scale, rates and means of economical
> production. The innovative emphasis is on products; funding is by
> private risk capital; and the environment tends to be close for pro-
> prietary reasons and tightly managed. Essentially all such work takes
> place in commercial laboratories.

Kennedy perceives that this three-stage process of innovation
is now being compressed in a revolutionary way. He describes
this compression as resulting from a fundamental rearrangement
of the social sponsorship of discovery to which several forces
contribute:

> First, a number of scientific disciplines are now being recognized as
> "ready" for accelerated application. As a discipline matures in power and
> confidence, leaps from the laboratory to applications that once seemed
> intimidating become commonplace. This now appears to be the case,
> for example, in immunology and genetic engineering, as well as micro-
> electronics.
>
> Second, there is a growing social awareness of the importance of scientific
> discovery to national productivity, and a consequent impatience with
> the traditional time requirements for diffusing technology to the public.
> In the past decade, various studies—particularly for biomedical research
> —have demonstrated that the typical time lag between the initial research
> discovery and practical application is ten years or more.
>
> Third, there is increasing concern in research universities, where more
> than two-thirds of the nation's basic science is done, about the retreat in
> public support for research. Federal funds for non-defense research have
> shrunk by 38% in real dollar value since 1968. Half of this decline took
> place in the first two years of this decade [the 1980s].
>
> Fourth, and perhaps most unexpected of all, the venture capital financing

of small, research intensive firms and fields like biotechnology and micro-electronics has been transformed. Since major changes were made in the capital gains tax, the investment funds available for such ventures have jumped from an estimated $70,000,000 in the mid-1970s to about $1.5 billion in 1982.

The result is an entirely novel mixture of influences upon university scientists and their institutions. For the University itself, there are new and challenging pressures on investment policy (Does the institution go into business with its own faculty?), on technology licensing (Should the University license inventions to faculty-led ventures? to their competitors? and if yes, under what terms? And should there be full disclosure of terms?), and on policies related to consulting faculty conflict of interest, and the protection of graduate student interest.

Many of the problems are simply not solvable by the institution alone. For the scientists themselves, and the "invisible colleges" that hold them together in national and international networks, there are other questions such as: How much can or should they guard against the withholding of information and exchange for proprietary reasons? How much involvement outside of faculty members' primary institutional affiliation is appropriate?

In general, this new climate offers more opportunities than it does problems. What we must try to do is involve industry more productively and creatively with university research in a way that leaves the latter intact, instead of risking fractionation of the training and research components and the division of faculty time between on- and off-campus ventures.

The University-Industry Connection

INDUSTRIAL SUPPORT OF UNIVERSITY RESEARCH

Data developed by the National Science Foundation (Figure 2) show that industrial support of academic research has been modest. In terms of the percentage of industrial research support in relation to total academic research expenditures, there is a sharp decline from 6 percent in 1960 to 3 percent in 1970, resulting from the rapid increase in federal support of university research and the relatively low amount of corporate support. An informal 1983 survey of major research universities shows that the percentage of industrial support of academic research for fiscal year 1980 was estimated at $235 million.

Overall, industry performs a fairly constant 70 percent of all U.S. research and development. But between 1960 and 1970, the percentage of this total directed toward basic research by and in

Fig. 2. Two Ways of Looking at the University-Industry Connection

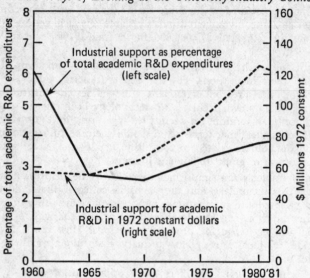

industrial laboratories shrank significantly, dropping from nearly one-third to about one-sixth of total basic research activity in the U.S.

Fig. 3. Industrially Supported Academic Research as a Percentage of Industrially Supported In-House Research—by Character of Work, 1960–81

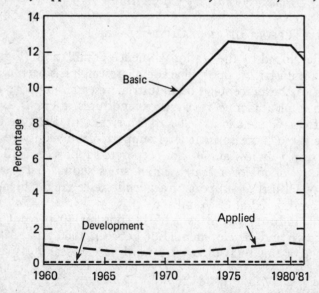

The portion of industry budgets allocated for support of university basic research increased from a low of about 6 percent in 1965, to a level of about 12 percent in 1974, where it essentially has remained (Figure 3). News media reports suggest industry sponsored research in universities tends to focus on a few fields. In 1979, nearly half of all industry sponsored research was within engineering, the largest percentage in chemical engineering. But industry does not interact with universities in innovation solely through contractual research.

Even a large percentage increase in the industry support (3.8 percent in 1981) would not have great effect on dependence by universities on federal research support (Figure 4). The overwhelming significance of federal support is even greater for universities without state funding, which includes many of the major U.S. research universities.

Fig. 4.   Sources of Support for Academic Research and Development, 1960–81

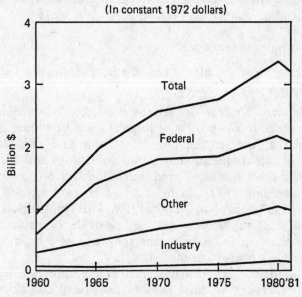

STAGES OF UNIVERSITY-INDUSTRY INTERACTION

Ties between a university and a company progress through several stages. At first a company may become aware of university technology and expertise useful to its business interests through an academic consultant or the university's technology licensing office.

The second, or "research" stage, derives from the interaction with the academic consultant or the person who provided the technology to the company from the university licensing office. In this stage, the academic, having gained a better understanding of the technology needs of the company, suggests a line of research to be conducted at the university.

The third, or the "application" stage, occurs when the company uses the research results (in some cases under license from the university), hires students, and engages the academic as a consultant to assist in adaptation of research results to their products and processes.

The fourth, or "philanthropy" stage, occurs when the company makes unrestricted gift support available to the university. This recognizes that alternative costs of research might have been substantially higher. Companies often support those areas of the university from which they draw most of their employees, including the liberal arts. Corporate matching of individual employee gifts to their alma maters has become very widespread.

UNIVERSITY-INDUSTRY LINKAGES

Eleven of the more prominent linkages between universities and industries are reviewed below.

The Graduated Student—By far the greatest contribution that universities make to the process of innovation is providing graduates qualified at the leading edge of science and engineering. There is growing competition between companies and the universities themselves for these graduates. Both have shortages of doctoral researchers in certain fields like computer science, electrical engineering, and plant biochemistry. This competition leads to the "seed corn" problem, where the loss of the best researchers from universities to industry means they will not be available to teach the next generation of students.

In some academic departments, such as computer science and electrical engineering, as many as 30 to 50 percent of all doctoral candidates are foreign students. In these fields, many American students go into industry after receiving a master's degree, which leaves foreigners comprising half of the doctoral recipients in the U.S. Most of them remain in the U.S., both to teach and to

join industry; the U.S. is eating the "seed corn" of other countries. Such students from developing countries are sorely needed
back in their homelands after they complete their training in the
U.S.

Another effect is realized in high-technology academic departments with large proportions of foreign graduate students.
Graduate students typically teach undergraduate sections, and
American-born students complain of great difficulty in understanding or communicating with many of their instructors.

Academic Consulting—Opportunities for consulting differ considerably by academic field (Figure 5). In 1969 nearly 66 percent
of academic engineers reported paid consulting activities, as compared to less than 33 percent of their physical and biological

*Fig. 5.   Faculty Participation in Consulting for Pay, by Academic Subject
Field, 1969*

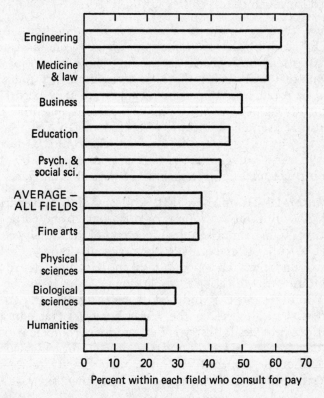

Percent within each field who consult for pay

science colleagues. While about 50 percent of the faculty in the professional schools reported paid consulting, only 20 percent of the humanities professors were so engaged. While recent data are not available, it can be perceived that the percentage of faculty in the biological sciences engaged in consulting will have increased substantially.

The president of Genex Corporation has pointed out that in 1978 there were only 4 companies worldwide that specialized in recombinant DNA technology, with a total capitalization of roughly $20 million. By late 1981 there were 110 recombinant technology firms with about $700 million capitalization. In addition perhaps 120 companies worldwide are currently in recombinant DNA technology. Since there is insufficient in-house expertise, these companies are strongly dependent upon close collaboration with academic scientists. In time, the growing competence of biotechnology research in these companies will lessen the need for much of this collaboration.

University and Industry Research Collaborations—As the TRACES study illustrates, collaboration between industry and universities may be required to produce those revolutionary innovations that will enable the U.S. to maintain its competitive posture in commerce. Important changes are now occurring in science and engineering which have enormous potential payoffs in industrial use. These include recombinant DNA research and solid state physics as it applies to microelectronics. Other areas that have been less glamorous and perhaps less visible to the public include materials research and artificial intelligence.

Philanthropy—During 1980–81, colleges and universities reported $778 million in voluntary donations from corporations (Figure 6). This comprised 18.4 percent of the total voluntary support to colleges and universities from all sources, including alumni foundations and nonalumni individuals. Contributions from industry to educational institutions can be both charitable and for self-interest. U.S. industry benefits significantly from the trained students, as well as the research results that educational institutions provide. Industry is in a uniquely competent position to evaluate institutions and university projects for which contributions are sought, generally in areas that directly relate to the commercial interests of a company. This may skew corporate

Fig. 6.   *National Estimate of Corporate Voluntary Support of Colleges and Universities, 1974–75 to 1980–81*

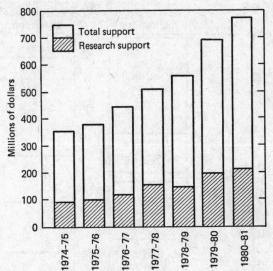

gifts away from the humanities, but it does free unrestricted university funds from the more technical departments in order to support the humanities. The data in Figure 7 exclude some of the largest donors of corporate philanthropy such as IBM and DuPont, which make their gifts directly rather than through company sponsored foundations.

Industry Affiliates Programs—Industry affiliates programs provide a channel of convenient and direct communication between university faculty and graduate students and member company scientists and engineers. Access to students is considered one of the prime reasons that companies, through an annual membership fee, join affiliate programs. Annual symposia on campus give company representatives an opportunity to both learn of current research and gain first-hand knowledge of the nature of research conducted by graduate students. Affiliate programs also encourage campus participation by scientists and engineers from member companies in seminars, colloquiums, and other campus activities. Visits by faculty members to affiliate companies may give both a chance to learn more of each other's research concerns. Industry affiliates are encouraged to bring technical problems of a nonproprietary nature to the attention of faculty members. This

Fig. 7. *1979 Grants of Endowed and Company Sponsored Foundations for Science and Technology, by Field*

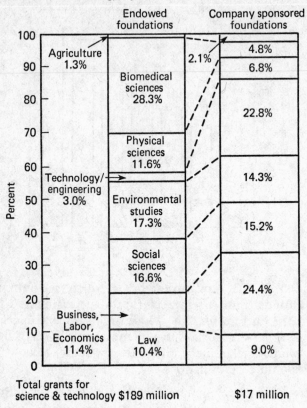

Total grants for
science & technology $189 million $17 million

may influence research directions at the university. Affiliate companies are usually provided early access to reports and publications in their area of interest, as well as the resumes of students.

Research Consortia—Research consortia, in contrast to industrial affiliate programs, are created to address specific mission oriented research when economies of scale are such that fragmented industry and university research is less likely to enable national industry to meet organized foreign competition. In fact, the U.S. Department of Justice has issued guidelines relaxing antitrust strictures, thereby enabling and encouraging collaborations involving many competing companies.

Financing for such consortia generally is a mix of corporate philanthropy, corporate research support, and federal research support. An example would be the Stanford University Center for Integrated Systems project that explores microelectronics (in particular, very large-scale integration [VLSI] microelectronic circuits). Such consortia may be stimulated by the highly publicized collaboration of Japanese government and industry in "target" technologies. These Japanese efforts both trained people for industry and provided the critical mass for new scientific and technical developments in the targeted technology.

Publications and Conferences—A free and open flow of ideas from universities to industry results from the swift publication in journals, scientific meetings, and conferences of the most current research results.

Scientist Exchanges—Definitive data are not available as to the nature and quantity of temporary appointments of company scientists to universities and of university scientists to companies. Judging from a 1983 discussion of university-industry interactions between the author and a group of German university presidents, this practice is much more prevalent in Germany than in the U.S.

Shared Equipment Use—Opportunities for collaborative use of expensive research equipment are often underexploited. There are several reasons for this. One is the proprietary nature of industry research. Another is the owner's priority for access to the equipment. In addition, universities seeking to make their research equipment available to industry and to share the equipment maintenance costs by charging for such access may be in jeopardy of violation of their nonprofit status.

The university can find itself in unfair competition with private companies that do not operate in a tax-free mode and are in the business of renting or leasing specialized research equipment. The NSF has certain guidelines for determining which NSF funded specialized research equipment at universities can be made available to industry in order to avoid such unfair competition. Clearly, if the research equipment is unique, there would not be a question of unfair competition.

A common organizational arrangement for access by industry

to equipment, as well as to consulting services, is utilized by universities in the United Kingdom. An entity, separately incorporated, is established either on university grounds or conveniently adjacent to university grounds. This legal entity acquires the specialized research equipment and makes access available to industry. These entities often also act as agents for faculty consulting, typically adding a surcharge on the order of 10 percent to cover their effort in arranging such consultantships. Such entities also provide a locus for more applied research which may not be appropriate for academic departments.

Industrial Parks—To encourage close interaction of industry and universities and to facilitate local innovation, many universities or communities seek to establish research parks in close proximity to the university, such as Research Triangle in North Carolina. While there is considerable momentum in the U.S. to establish such industrial parks, only a few universities have been able to achieve any success. In summarizing its study of three forms of university-industry collaboration (research parks, cooperative research centers, and industrial extension services), the General Accounting Office claims ". . . the most dramatic contribution to innovation appears to be made by research parks."

Technology Licensing—Since the mid–1970s there has been a significant growth of on-campus university technology licensing departments. This is illustrated by the membership of the Society of University Patent Administrators (SUPA). At the end of 1975, the year of its first annual meeting, SUPA had 51 members; at its 1983 annual meeting, 226 individuals attended, and membership growth continues to increase. This development reflects desire of universities to establish their own technology licensing programs in contrast to using separate patent management organizations. It often is more economical for a university to use a patent management organization until its research volume reaches a stage where an on-campus organization can be justified.

Research Corporation, a nonprofit patent management organization, was established in 1912 based upon patents governing the electrostatic precipitator donated by Frederick Gardner Cottrell, then professor of chemistry at the University of California at Berkeley. Research Corporation retains a percentage of gross royalty income and utilizes such revenues in a program of research grants which total more than $60 million to date. Such "seed money" grants, usually to beginning scientific investigators, have been of great value, often leading to the establishment of major academic research programs for which continuing funding of larger amounts is obtained from federal research agencies such as the National Science Foundation.

Universities typically share between 15 percent and 50 percent of royalty income with inventors. At many universities all inventions of university staff, faculty, and students are required to be assigned to the university; at other universities, only those inventions which occur under sponsored grants and contracts are assigned to the university. However, because university inventions normally are undeveloped, requiring significant risk capital to develop a marketable product or process, a university typically must grant an exclusive license (for a limited exclusive period) to a company in order to encourage such investment. After this exclusive period, the intellectual property under the license is made available on a nonexclusive basis to all companies. Public Law 96–517 provides that first option to an exclusive license to a U.S. patent arising from federally funded research must be for U.S. manufacture.

The oldest university technology licensing program appears to be that of the Wisconsin Alumni Research Foundation (WARF) established in 1925 to exploit the patents of Professor Harry Steenbock for the benefit of the university. Through both royalty revenues and shrewd investment of them, WARF has given over $100 million to the University of Wisconsin. Annual donations have averaged about 5 percent of the university's research budget, and it has been suggested that this research funding has been a significant factor in the eminence of its research program, providing the important leverage of "free" research dollars without the extensive administration involved in proposal preparation, reporting, and other "strings" of federal and industrial research support.

The amount of direct license income (excepting any income from investments derived from such royalty revenues) has not been large at U.S. universities. During 1981–82, less than ten universities received more than $1 million in royalty revenues; the largest amount received was $2.5 million. Although greatly increased emphasis on technology licensing and university-industry interactions may cause royalty revenues to grow substantially in future years, technology licensing programs tend to have a greater influence on universities through establishment of industrial linkages than in direct royalty revenues.

FOSTERING UNIVERSITY-INDUSTRY INTERACTIONS

In general, the ease of university and industry interactions in the United States is looked at with envy by other countries, often singled out as a model for their own future growth. The interaction has stemmed more from the initiatives by the universities and industries than from the government. But the sustained, indirect involvement of government through its support of basic research at universities has enabled them to train students and foster innovation by industry. Increased university-industry research collaboration has been widely forecast for the 1980s. As the NSB report notes:

> Questions are raised about whether industry has sufficient resources available to increase allotments to university research; whether academic research can really benefit industry; whether academic freedom and openness of scientific communications can be preserved in the face of the constraints and temptations of commercial enterprises. But the new arrangements highlighted here reflect an optimistic mood that is grounded in an awareness that the problems and opportunities in technologically based industrial production are substantially different from those in the past.

The NSB report suggests three factors that characterize the present situation.

The first factor is that product and process improvement in innovation in some industries has evolved to levels of complexity that demand understanding of fundamental physical and biological phenomena, thereby requiring much higher levels of training in and use of basic science in engineering than the "cut-and-try" inventor of yore.

The second factor considers that incremental advances in narrow technical areas, which may have been characteristic of much industrial development in the past, are giving way to use of a broad range of science and engineering disciplines on complex, often ill-defined, problems or exploitations of new analytical capabilities. Hence, it is becoming increasingly difficult for any single industrial laboratory to fully encompass the required expertise. The

NSB report suggests that a partial remedy may be for industry to seek out "the pertinent skills" in the nation's universities.

The third factor notes that the rapid expansion of the nation's research and development system following World War II "has diffused research capabilities over a much broader range of institutions—academic and industrial—than before." This suggests the future unlikeliness that any single company can hold and maintain a leading edge on technical advance in a given area, such as Du-Pont's experience in polymer fibers.

The Challenges Ahead

In general, while there are certainly areas for improvement, the linkages between government, universities, and industry work extremely well, but there is no basis for complacency as competition is rapidly closing the gap. This is evidenced by the declining competitiveness of the United States in many market areas.

Productivity has been dropping in the U.S. since 1978, and our share of the world's market declined by 23 percent in the 1970s. In high-technology goods, the United States' share of the world market declined from 30 percent in the 1960s to about 20 percent by 1982. Selected industries in high technology showed even sharper percentage drops: telecommunications fell from 30 percent to 19 percent, scientific instruments declined from 30 percent to 15 percent, and pharmaceutical drugs decreased from 28 percent to 15 percent.

Egils Milbergs, director of the Office of Productivity, Technology, and Innovation of the U.S. Department of Commerce, perceives the following five basic "forces" where government policy is needed to accommodate the challenge of international competitiveness.

Industrial Targeting Strategies—This is illustrated by the Japanese, whose industrial targeting strategies have brought new products to the market much faster and with a much higher quality and reliability than U.S. firms have been able to do. Governmental actions to counter this competition include direct funding of research and development projects, preferential access to procurement, import protection, and other such measures.

Newly Industrializing Countries—Countries such as Mexico,

Brazil, Saudi Arabia, Korea, and Singapore are expected to join in the competition for new markets in a large way. Competition from these countries is already beginning to affect Japan in areas such as steel and automobiles where U.S. competitiveness had been significantly eroded earlier.

Increased Rate of Technology Change—The rate of technology change acts to accelerate the obsolescence of plants and equipment. For example, the lifetime of research equipment twenty years ago was fifteen years, whereas in the 1980s, the lifetime is four years.

Changing Demographics—This fourth force for change is U.S. human resources. Milbergs notes that the new labor force has a higher expectation from the work environment, desiring to share more in management decisions and profits. Emphasis is being placed on more benefits and fewer hours. Dislocations are anticipated because of shortages of technically skilled individuals in key technology areas and pools of displaced workers in other technology areas. Milbergs observes:

> It is possible that by the year 2000, over half of the labor force in the manufacturing sector will be replaced because of automation, rationalization, foreign outsourcing, or the fact that we no longer have a comparative advantage in a particular manufacturing sector.

Change in Management Philosophy—Present U.S. industrial management is under sharp criticism for the emphasis of short-term results rather than long-term, more strategic investment. Another manifestation of this management system is the plethora of adversarial proceedings, one aspect of our society that other countries do not desire to emulate. To be a Master of Business Administration, Doctor of Medicine, or Bachelor of Laws has long been more prestigious among youth in our society than to be a chemist or engineer, yet these latter professions produce the products and services on which industry is based and which positively influence innovation.

Cecily Cannan Selby

6

Current Trends in Mathematics, Science, and Technology Education:

Implications for Technological Innovation

Introduction

Identification and encouragement of innovative thinking and practice and of technological understanding as educational objectives are notable omissions in all but a very few of the plethora of articles, studies, reports, and recommendations about elementary, secondary, and college education which mark our current time. At a time when our leaders of government, industry, and academe are extolling the crucial value of innovation in scientific and technological endeavors and when vast improvement is being called for throughout all of education in these areas, this is a puzzling omission. In the watershed of interest in and concern about schooling in general, especially precollege education in mathematics, science, and technology, the case is repeatedly made

CECILY C. SELBY *was a cochair of the National Science Board Commission on Precollege Education in Science, Mathematics, and Technology. Her broad professional career has encompassed the fields of science, education, management, and communications. A member of the board of RCA Corporation, the National Broadcasting Corporation, Avon Products, and until recently, Loehmann's Inc., as well as a trustee of Radcliffe College and the Brooklyn Law School, Dr. Selby is, in addition, a member of the Rockefeller University Council, the Corporation of the Woods Hole Oceanographic Institution, and formerly of the Corporation of the Massachusetts Institute of Technology.*

that technological advances require scientific and technological literacy in the total population, new skills in the work force, and an expanding pool of future scientists, engineers, and technologists. Students in this pool must be capable of both the solid achievement and the innovations that can lead the country's technological (and thus economic) advances.

Beyond one valuable report *(Learning Environments for Innovation,* U.S. Department of Commerce, 1980), little is said about preparing students at either the elementary, secondary, or college level for *innovative thinking and working.* This may reflect a resistance to teaching toward an objective that cannot be measured or one of several other assumptions: that such talent is too rare to be worth a concerted effort to develop, that we do not know how to develop it, or that it is really not that important. Alternatively, it could mean that the problems faced in moving our entire school population a giant step ahead are so monumental that issues of individual creativity appear to decision makers to be of much lower priority. These assumptions must now be seriously challenged.

NEEDS OF A TECHNOLOGICALLY DRIVEN SOCIETY

Responsible leaders in all sectors recognize that a technologically driven society requires some degree of scientific and technological literacy for all who would live productively within it. Education for appropriate understanding and skill in mathematics, science, and technology must thus move to share center stage with the other liberal arts throughout all of schooling. What instructional objectives should be included in mathematics and sciences and be available to all students? What do we mean by technological literacy? Objectives usually mentioned include the ability to solve problems, to master appropriate subject matter, and to approach issues with rigor and the ability to quantify and analyze. Should they also include encouragement, or even legitimization, of some students' interest in and ability to deal with ambiguity, to take risks in thinking or in extrapolation from observations, or to explore radical, extreme alternative hypotheses for problems that are posed?

That such objectives are not addressed in most current educational planning, particularly at the precollege level, is understandable when one considers the historical trends which led to the educational policies and practices of today. Neither innovation nor technology has yet been considered within overall goals for educational planning, for teacher training, for school organization, or for precollege curriculum development. With respect to innovation, is it that we consider the role of public education primarily one of socialization which should reward conformity in thinking and behavior, encourage allegiance to hierarchical organizations (such as the traditional student government), and take satisfaction with the body of knowledge being communicated? Have we assumed that we need only the few "elite" innovative and creative thinkers who would emerge or would be cultivated by the more privileged educational system or opportunities such as science fairs and talent searches outside formal education? With respect to technology, have we considered it primarily an issue of job training or vocational education? Is it again a subject for out-of-school learning up to the level of preprofessional education?

It will be instructive to ask these questions in the light of the historical and politico-sociological trends that brought us to today and then to consider changes in educational objectives required by current conditions and future needs—particularly those related to technological development and innovation.

Historical Review: How Did We Get to Where We Are?

Before Thomas Jefferson's leadership in committing the United States to free public education, privately funded institutions for education (such as William and Mary [1693], Harvard [1636], and the various academies of New England, including the Boston Latin School [1635]) were established in the colonies. These schools and colleges were direct descendants of the aristocratic "liberal education" of England and Europe, stressing the classics, literature, mathematics, and natural philosophy (science). In the U.S. the Ordinance of 1785 set aside public lands for the support of schools in every township, proclaiming that "schools and the means of education shall forever be encouraged." Initially the program at public secondary and elementary schools followed the classical tradition but included some practical skills following Benjamin Franklin's recommendations regarding "useful learning."

During the late 1820s in England, a reaction arose to discrepancies between the quality of most elite private grammar ("public") schools and those available to others for lower fees. The leadership of this movement came from the middle class, whose income was derived largely from commerce and industry. They sought more utilitarian ends for their students and founded schools managed by a committee, proprietors, or a managing board (in today's language) with an emphasis not so much on producing gentlemen, but rather individuals for industry, commerce, and the services. While the curriculum remained classically based, it included more "modern" subjects and much greater emphasis on mathematics. Often schools were organized for students over the age of fourteen into classical and modern divisions. According to Geoffrey Howson *(A History of Mathematics Education),* this movement led to considerable interest in and attention to the methods and rationale for the teaching of mathematics in the United Kingdom which influenced developments elsewhere.

Particular problems of teaching and learning mathematics gradually became more explicitly and professionally scrutinized. For example, in 1836 Augustus De Morgan, writing on the goals of mathematics, stated that it was not sufficient to justify mathematics a place in the school curriculum because it is useful. He argued that law, medicine, and architecture are also useful but are specialized subjects to be embarked upon only once a general education has been completed. He, and most educators following him, saw the principal contribution of mathematics to general education as a vehicle for the enhancement of the faculty of reasoning. De Morgan addressed the dual aspects of mathematics —the practical and the contemplative (an important continuing consideration as one deals with this subject):

> The actual quantity of mathematics acquired . . . is . . . of little importance, when compared to the manner in which it has been studied, at least as far as the great end, the improvement of the reasoning powers, is concerned. We might be tempted to say, let everyone learn much and

well; well in order that the habits of mind acquired may be such as to act beneficially on other pursuits; much in order to apply the results to mechanics, astronomy, optics [etc.] which can never be completely understood without them.

In U.S. developments of about the same era, the initial concept of the liberal arts on which the early institutions were founded was picked up within the Jeffersonian education philosophy: the principle of free higher education for those who have the talent and motivation to benefit from it. This became accepted political philosophy with the Ordinance of 1785 for schools and the founding of the University of Virginia (1819) as a public state funded college.

UTILITY AND EDUCATION

The next trend in U.S. education, unique in its pervasiveness in the Western tradition, was the Jacksonian emphasis on utilitarian ends. Such objectives for education became reality with the founding of land-grant colleges for agriculture and the mechanical arts under the Morrill Act of 1862. Elementary and secondary schools participated in this utilitarian vocational thrust by means of the Smith-Hughes Act of 1917 and, later, The Vocational Education Act of 1963.

During the latter part of the nineteenth century, the primary obligation of educational institutions was perceived to be to provide students with the skills and attitudes that would allow them to perform the tasks the society needed. When Justin Morrill, Republican representative from Vermont, introduced his legislation in 1856, his intent was for students from each congressional district to receive a scientific and practical education at public expense. He believed the nation needed this new expertise to increase its productivity and found that existing colleges were little interested in providing instruction in subjects such as science, agriculture, and mechanics. Apparently Morrill recognized the potential benefits to *individuals* of state supported, low-tuition colleges, but these advantages were inadequate to persuade his colleagues to pass his original bill. His bill had considerable opposition, taking six years from introduction to passage. By this

time, amendments to the legislation provided for federal land
grants to each state to establish universities providing instruction
in agriculture, the mechanical arts, and training for military
officers. Morrill's argument, although unsuccessful in the late nineteenth century, would prove to be successful in gaining adherents
in the twentieth century when equal education opportunity became a profound educational goal.

According to reviews by Patricia Albjerg Graham, the variously
inspired efforts from 1862 to 1914 to provide federal aid to higher
education had one unifying theme—that the product of that education, whether it be research, demonstrations, or instructed students, would be valuable to the United States in terms of improved
industry and agriculture. In 1870 Calvin Woodward, a Harvard
mathematician, complained that schools were training students to
be "gentlemen" rather than preparing them for work.

The parallel development of the land-grant colleges of the nineteenth century and the new emphasis on research in U.S. universities later in that century (e.g., at Johns Hopkins and Clark)
continued the side-by-side development of utilitarian and intellectual liberal arts approaches and was successful in strengthening
both the intellectual and technological base of the American
economy and society. Indeed, for the most part, the nation
retained its confidence in the overall system until after World
War II.

Harvard had instituted the Bachelor of Science degree in 1851
to distinguish between completion of a program focused on
modern scientific subjects (by omitting classical studies) and
completion of a traditional liberal arts program grounded in the
classics. At Bowdoin a comparable distinction was made by
whether or not Greek and Latin were offered for entrance. By
the beginning of this century at Harvard and elsewhere, an
elective system of courses was introduced, pushing out the old
classical model. Distribution requirements were then added and
organized by departments to try to maintain some sense of a
required core and a stable program. Leadership in redefining
what such a core should be was provided by Columbia University.
Based upon its World War I experience in educating officer candidates on the background of the conflict, Columbia College, in

1923, developed and introduced its two-year sequence called "Contemporary Civilization," which served as a model for programs in general education later introduced in the 1930s and 1940s (e.g., Harvard's "Report on General Education" in the late 1940s).

CREATIVITY AND EDUCATION

Starting in Europe and England with Comenius, Rousseau, Spencer, and Froebel and continuing in America with the leadership of John Dewey, questions about how people learn—and, therefore, how to teach them—gave birth to the progressive movement in education of the 1930s. The major educational trend developed several significant independent schools and affected teacher training, individual teacher initiatives, and movements such as that of the "open classroom." Emphasis on the individual needs of and the creativity inherent in each child led to discovery-and-inquiry methods of teaching, individualized instruction, and independent study.

At the beginning of the twentieth century, individual educational outcomes were beginning to take precedence over societal ones, initially for the children of the well-to-do. Partly owing to the influence of the progressive movement, some educators were beginning to believe—and to argue—that their primary obligation was to the child and not to society. Perhaps another reason why many educators in the first half of the twentieth century were willing to shift focus from the society to the child was, as Graham suggests, because of their changing view of American society. If one believed that America had accomplished the massive initial tasks it faced as a nation—conquering frontiers, assimilating immigrants, and becoming accepted as a world power —then perhaps it could afford to concentrate on the needs of its children and on unleashing their creativity.

The attempt to enhance creativity and the effort to increase educational opportunity were luxuries that many saw the nation could not afford when the energies of its citizens were required for the more pressing tasks of gaining preeminence in the world. This point will be important to remember as we look at the thrust of most educational recommendations being made in the late twentieth century with emergent concern about U.S. preeminence as a world power.

The educational issues we are facing today thus arise from a tension among the four educational approaches that brought us to this point: liberal arts intellectualism, Jeffersonian egalitarianism, Jacksonian utilitarianism, and the student-centered developmental approach of progressivism. If the inclusion of creative thought and action in educational goals is a luxury for the few, then how can the many have true access to fields such as science, mathematics, and engineering where the introduction of concepts and processes associated with "elite" education at an early age can be shown to be the only true access? How can we do justice to the extraordinary variety of cultural backgrounds of students in our precollege and college systems in an education (including technology) for useful participation in society and also provide access to opportunities for the highest level of intellectual, innovative, and creative endeavor within the fields of mathematics and science? How can we keep children's own interests and talents alive throughout a "basic" education considered necessary to provide them with skills that contribute to society? How can schools help children retain their individuality and integrity and yet prepare them to live in an industrial society requiring conformity without being either alienated or crushed by it?

The Current Status of U.S. Education: Where Are We?

The 1983 report of the National Commission on Excellence summarizes well the depth and breadth of concern about current school and college conditions:

> Our nation is at risk. Our once unchallenged preeminence in commerce, industry, science, and technological innovation is being overtaken by competitors throughout the world. . . . We report to the American people that while we can take justifiable pride in what our schools and colleges have historically accomplished and contributed to the United States and the well-being of its people, the educational foundations of our society are presently being eroded by a rising tide of mediocrity that threatens our very future as a nation and as a people.

Focusing particularly on elementary and secondary education in mathematics, science, and technology, the National Science

Board (NSB) Commission on Precollege Education in Mathematics, Science, and Technology reported later in 1983:

> Alarming numbers of young Americans are ill-equipped to work in, contribute to, profit from, and enjoy our increasingly technological society. Far too many young Americans have emerged from the nation's elementary and secondary schools with an inadequate grounding in mathematics, science, and technology. . . . At a time when America's national security, economic well-being, and world leadership increasingly depend on mathematics, science, and technology, the nation faces serious declines in skills and understanding in these areas among all our youth.

The concerns are comprehensive, encompassing issues of general literacy, including science and technology, as well as the recruitment and education of our future scientists and technologists. While the nation still takes justifiable pride in the education of its top students, the question is whether we are maintaining a large enough pool of students able to lead our scientific and technological endeavors in the future. To maximize this nation's resources, as much as to promote educational equity, there must be a much greater recruitment into this pool of students who are female, from minority groups, and/or socio-economically disadvantaged.

What is the underlying pathology that has led to decreased student achievement and participation in mathematics and science, just at a time when both the needs of the nation and of its individual citizens require otherwise?

The evidence is that students enter school in the primary grades with an interest in numbers, in spatial relationships, and certainly in the world around them but are "turned off" from the study of mathematics and science in the early grades and generally discouraged from continued study in these subjects. Children with encouragement from home and out-of-school opportunities to pursue these subjects (as will be discussed later) tend to be those electing to study them further. Clearly, middle- and upper-class white males predominate in such a population. In the secondary school, election of advanced science courses is usually by the already motivated, college preparatory, preprofessional student. Future career choices and the mathematics electives that go with them are generally made around the eighth and ninth grades, when disenchantment with mathematics and science is prevalent and encouragement most needed. Minorities, females, and the socially disadvantaged are generally lacking appropriate counseling and encouragement and consequently are underrepresented in college preparatory science and mathematics courses, as well as among science and engineering majors and in careers.

Dropouts among college science majors include some of the most creative students and a high percentage of females and minorities. Graduates electing doctoral work and electing university teaching in science and engineering are not sufficient in number to meet the nation's projected needs for research and advanced teaching. The number electing elementary and secondary school teaching in mathematics and science most certainly are not nearly adequate to meet the current and projected needs of our school population. Most states now report acute shortages of physics and chemistry teachers, and science and mathematics teachers are leaving the profession at a much faster rate than ever—primarily for jobs in industry.

Most secondary schools' science courses are designed for the college preparatory, preprofessional student and make little, if any, attempt to relate science to society and the individual, let alone relate each of the sciences to each other. College courses remain discipline centered with few real efforts at interdisciplinary or technology related approaches successful or capable of replication from one faculty to another. University courses of study and departmental offerings became increasingly fragmented after World War II under faculty members who became more research oriented in response to spectacular increases in available funding. New technologies and advances in science led to a change of thrust in many of the sciences, with introductory science courses often left behind and unchanged. Science courses for the nonscientist continued to be remarkably unsuccessful. Of course, notable exceptions to these generalities can be cited, but it is the overall situation that concerns us.

Perhaps the most promising development in undergraduate science and teaching has been involvement of undergraduates in research. Liberal arts colleges led the way, but the large research universities are now increasingly providing such opportunities and extending them to high school students. Such programs have been notably successful in encouraging students to continue in science and to develop creative and innovative approaches.

GOAL CONFUSION AND EXTRANEOUS FORCES
IN HIGHER EDUCATION

The American Assembly's final report on *The Integrity of Higher Education* (1979) states:

... public confidence in [American higher education] has been eroded in recent years. Consensus on what constitutes legitimate higher education has been reduced and expectations of it—and claims for it—have not been fulfilled. ...

Dissatisfaction with college education in general can be traced to changes following World War II. According to the 1983 publication of the American Association of Colleges Project on Redefining the Meaning and Purpose of Baccalaureate Degrees entitled *A Search for Quality and Coherence in Baccalaureate Education,* American colleges have now created over 500 different baccalaureate degrees with little in common and with widely different standards. Clearly a desire to maintain enrollments, as the college-age population declined, has motivated the introduction of such degree programs and has caused some changes in standards. Public loss of confidence in what a college degree stands for—what degree of literacy, skill, or understanding it can be relied upon to provide—has much justification.

We continue to be in the middle of a confusion rather than a congruence of the historic trends summarized earlier.

Such confusion in our objectives also leads to the phenomenon of the "hidden curriculum," as articulated by Dr. Benson Snyder at M.I.T. There is often an incongruence between the messages coming to students from the formally stated goals of teachers and the curriculum and the means that students must use to attain high grades and other academic recognition. Alienation, hopelessness, the dropping out by the most creative of students as well as the most hopeless (albeit for different reasons), and much of the unrest that schools and colleges lived through in the 1960s and 1970s can be traced to educators' failure to recognize the discrepancy between what they were preaching and what the "hidden curriculum" of their values and practices required of students. "Don't do as I do, do as I say" is perhaps the most honest cry of parents, teachers, and professors alike!

POSSIBLE CAUSES OF DISAFFECTION

Before considering recommendations for action to turn back and redirect this tide now running against mathematics and science in our schooling, it is important to try to understand its causes. Rather than the *symptoms* that every study, report, and journalist finds temptingly easy to list, it is the underlying pathology and its elusive diagnosis and therapy that must be the focus for those who seek to address the symptoms. What has been going on in our nation's classrooms and lecture halls to discourage students and their teachers from further participation in these fields? Studies funded by the National Science Foundation and carried out in the late 1970s and early 1980s by the National Science Teachers Association, the American Chemical Society, the American Association for the Advancement of Science, and other professional organizations provide comprehensive and in-depth data and analyses. A synthesis of all the reports and observations reveals the following characteristics of most elementary and secondary science instruction:

1. the supremacy of the textbook—90 percent of all science teachers in the U.S. use a textbook 90 percent of the time and attempt to cover all the book's content;
2. the implicit justification of course content as preparation for the next level—not as a response to student need, interests, or skill level;
3. limitation of the goals of science instruction to certain specific knowledges and practices—rather than recognition of the multidimensional scope of science;
4. continued compartmentalization of science in the order of presentation of the various disciplines—rather than consideration of the *total* interdisciplinary science curriculum as a dynamic relationship of materials, teacher, and environment;
5. the predominance of the lecture format, the teacher talking and maintaining control—an approach effective only for students who want to succeed in traditional ways; and
6. laboratory exercises that merely require following directions and verifying information given by the text or the teacher—most science courses do not include a single laboratory experiment where students can identify and define a problem and participate in any decisions about procedures, observations, and interpretation.

With success in school science deriving from emphasis upon content for its own sake, the teacher's lectures, and textbook presentations, the students who achieve in this context generally experience great disillusionments later (unless they continue along the same track in college) when they find that this kind of science (which is not science!) is not of use to them in later life.

John Goodlad, in reporting in 1983 on an eight-year, in-depth study of 1,016 classrooms across the country, found the same phenomenon in all classrooms in all subjects: teachers appear to teach within a very limited repertoire of pedagogical alternatives emphasizing teacher talk and the monitoring of seatwork. Customary pedagogy places the teacher very much in control. Feedback-with-guidance associated with helping students to understand and correct their mistakes is rarely found. The result is a numbing boredom and alienation of students—and a failure to grow in learning and in the ability and confidence to solve problems. "Rarely did we observe laughter, anger, or any overt display of feelings," Goodlad comments. "If I myself were in such classrooms hour after hour, I would end up putting my mind in some kind of 'hold' position, which is exactly what students do." Most of us undoubtedly have done the same.

New Educational Objectives Needed: Where Should We Be Going?

The current status of the nation's educational practices indicates that, at a time when we would like our *total* population to have some skills and understanding in areas of mathematics, science, and technology, we have been most successful in discouraging general interest and achievement in these subjects. At the same time, although we would like to ensure a continuing supply of innovative thinkers and workers for our scientific and engineering (and also teaching) endeavors, we have been leaving the development of such talent to chance cultural and educational advantage and to out-of-school (informal education) opportunities. In fairness to all who have been laboring so hard in these vineyards, the objectives we now identify as in the best interest of the nation and of its citizens are new; that the old ways cannot meet new objectives should not be surprising. To say that mathematics and science must move from the periphery of learning for all but a few to center stage for everyone represents change and requires consequent change. Indeed, once we can agree on the new educational goals and objectives, the routes which must be taken to reach them will be clear—and become possible.

The National Science Board commission report cited above recommends that the nation's educational systems, both formal and informal, should have the capacity:

1. to continue to develop and broaden the pool of students who are well prepared and highly motivated for advanced careers in mathematics, science, and engineering;
2. to widen the range and increase the quality of educational offerings in mathematics, science, and technology at all grade levels so that more students would be prepared for, and thus have greater options to choose among, technically oriented careers and professions; and
3. to increase the general literacy in mathematics, science, and technology of all citizens for life, work, and full participation in the society of the future.

All these goals require new objectives for mathematics and science education and the addition of technology education—a newcomer to the liberal arts tradition.

New goals for the nation require new commitments: that the understanding of mathematics and science is not only possible but also a benefit for all, *and* that the pool of future leaders and talent in these fields must include those for whom a challenging education has not previously been provided. There appears to be general agreement that a changing technological society requires youth who are "trainable" and well prepared for further education in industrial and other sectors of the economy. Focusing on specific job related skills rather than on a general education at the elementary and secondary level is deemed ill advised by most of those reporting on the nation's needs in its work force.

Fundamental changes in instruction are required in the mathematics and sciences, not only to reverse declining student participation and achievement and to deal with new national goals of technological and scientific literacy, but *also* to adapt to:

1. the explosion of scientific and technical knowledge and concomitant change in judgment about what students should learn in each discipline;
2. the escalating availability of most effective interactive educational technological aids, particularly computers; and

3. advances in the cognitive and behavioral sciences in understanding how people learn and how such learning can improve teacher training, curriculum, and software development.

Acceptance of mathematics and science as additional windows on learning and growing, deserving places within general literacy, requires fundamental change in teacher and administrator attitude. Mathematics must be seen, as suggested by De Morgan in the 1830s, as a way of thinking that opens doors to new knowledge in virtually every field of endeavor (art, music, business, social science, etc.) and that is essential for advanced understanding of all science and technology.

A report by the Conference Board of Mathematical Societies (1983) suggests that elementary mathematics instruction in this computer age should emphasize practical, real-life applications, informal mental arithmetic, estimation, and approximation rather than paper and pencil computation, even though comprehensive understanding of and facility with number facts and related processes are considered as important as ever. At the secondary level, finite (discrete) mathematics must now be included with the precalculus topics, and new approaches for both must be anticipated from the development of computer science and computer technology. The current curriculum must be streamlined, leaving way for important new topics, such as the use and understanding of computers. Undergraduate mathematics curricula must necessarily respond to changes in secondary school teaching and could also constructively affect precollege curricula by judicious change in requirements and course prerequisites. The development of courses for current undergraduates should include discrete and computer mathematics not covered in secondary school. Applications to other fields of study, including technology, should be emphasized as much in college as in school.

PREPARATION FOR TECHNOLOGY INNOVATION

With respect to new criteria for technology and science education, the NSB commission recommends, in part:

Students must be prepared to understand technological innovation, the productivity of technology, the impacts of the products of technology on the quality of life, and the need for critical evaluation of societal matters involving the consequences of technology. Further, the nature of scientific inquiry and observation presents frequent opportunities for experiencing success. Such inquiry does not require unique answers. Students can rightly and successfully report what they have seen and found. This type of experience should be encouraged.

The commission report lists recommended criteria for improving and changing instruction in the sciences. Above all, observation, student inquiry, and "hands-on" approaches must be encouraged. Teacher or classroom "coverage" of any prescribed amount of material is to defer to development of interest and skill in scientific observation and to motivation of understanding the results of this observation. Particular talents for innovative and creative thinking must be developed and enhanced, along with the capacity for problem solving, critical thinking, and knowledge useful for living, as well as for advanced study.

The commission report recommends that technology should be included in the curriculum of kindergarten through grade 12 as a topic integrating science, mathematics, and other fields of study—not as a separate subject in the curriculum. With the leadership of the Alfred P. Sloan, Jr., Foundation, some liberal arts colleges are now exploring ways to integrate understanding of technology into their curricula. One can anticipate that such efforts will escalate at the college level, as indeed they should.

Coupled with objectives for the development of student skills and understanding areas of mathematics, science, and technology must be a clarification of the essence of these subjects—what is the nature of the mathematics, science, and technology that should be understood? Study after study indicates that, through school and college (always, of course, with notable and most precious exceptions) we have been communicating science as a factual, difficult, textbook-bound subject governed by "known facts" and something rigidly defined as "The scientific method"—the keys to this kingdom being discipline centered courses which are as effective in locking students out as in inviting them in.

SCIENCE EDUCATION: A PERSONAL VIEW

In thinking about the need to turn around such perceptions, I am reminded of a prayer I was given by a fellow school head many years ago: "Help us from speaking those things which are not true or, being true, are not the whole truth or, being wholly true, are merciless."

It appears, wherever one turns in exploring science and mathematics education, that the formal curriculum has been merciless in excluding our children from consideration of the unknown in both the process and the product of the study of the natural world. We have taken the results of several centuries of creative, risk-taking investigation and thinking, organized it into a hierarchical framework, and presented this organized body of knowledge as symbolic of the search itself. My favorite definition of science comes from Gerald Edelman: "Science is imagination in the search for verifiable truth." Imagination and the search are so often missing in the teaching of science.

So many thoughtful persons, whether coming from the point of view of psychology (like Jerome Bruner), from biology and medicine (like Lewis Thomas), from education and computer science (like Seymour Papert), or from physics and chemistry (like Gerald Holton), independently suggest that the unknowns in math and science should be the takeoff point of science teaching, titillating imagination and motivating learning. Bruner wants "the schools, like life, to bring into light the tough predicaments that pupils are already beginning to recognize and make them part of the coin of discussion. . . . Do not ask whether children are ready. Nobody is ever ready until given a chance."

Thomas speaks of the essence of science:

> The endeavor is not, as is sometimes thought, a way of building a solid, indestructible body of immutable truth, fact laid precisely upon fact in the manner of twigs in an anthill. . . . Science is not like this at all: it keeps changing, shifting, revising, discovering that it was wrong and then heaving itself explosively apart to redesign everything. . . . It is a living thing, a celebration of human fallibility; at its very best it is rather like an embryo.

Papert speaks comparably of mathematics:

> Mathematical work, as scientific work, does not proceed along the narrow logical path of truth to truth, but bravely and gropingly follows deviations through the surrounding marshland of propositions which are neither simply and wholly true nor simply and wholly false.

Gerald Holton and others suggest that science is inaccessible to the nonscientist. The scientist works on ever-increasing levels of abstraction until eventually:

> there is a fundamental logical independence of the concepts from the sense experience. The concepts are not some distillation of the experi-

ences which anybody, using the *kind of logical reasoning one supposedly learns in school,* should sooner or later be able to trace. . . . On the contrary, the concepts themselves are freely formed, subject to the *a posteriori* usefulness of the whole structure when confronted with experience.

If one's learning went beyond logical reasoning at school, would science be more accessible? When synthesis of thought or observation does not proceed along lines that previously yielded solutions of like problems, the creative and talented mind steps back, takes all the pieces of the puzzle apart, and then finds ways to reassemble them in new ways that fit. Is this approach ever part of any but the *most* talented teacher's or professor's teaching methodology? Can it be? Should such approaches be incorporated into instructional design? Would this help science be more accessible?

Teachers frequently comment that students assigned to classes for "low achievers" will often give wonderfully practical and imaginative answers to an unconventional question. Students identified as gifted and talented will more frequently have a "single right answer" orientation. Children from classes identified as for the "gifted" make clear how they carry constant and pervasive anxiety with them—e.g., pressure from parents and teachers. They worry about school performance and grades and, consequently, identify learning with single, correct answers. Successful students become adept at dealing with the "hidden curriculum"—a requirement for their success. They learn that innovation is not rewarded—except in very rare situations.

SIGNIFICANT OMISSIONS

Until there is a broader understanding of how the best scientists, engineers, and mathematicians perceive their fields, and how important technology and creative and innovative thinking are to them, mathematics, science, and technology will still be communicated in ways guaranteed to "turn off" most humanistic and imaginative students and adults alike. That this point fails of appreciation, consider some representative comments excerpted from recent reports on the status of our schooling.

The Commission on Excellence report (April 1983), *A Nation at Risk,* has much to say about the meaning of "excellence" for both students and schools. While the report speaks of maximizing all students' talents, it does not make a special point of innovative and/or creative thinking or of technology. However, in listing the "tools at hand," the *"ingenuity* of our policymakers and scientists" is considered one of our assets.

In the College Board report (1983) on *Academic Preparation for College: What Students Need to Know and Be Able to Do,* there is an emphasis on what are called "basic academic competencies" (reading, writing, speaking and listening, mathematics, reasoning, and studying). Although the mastery of currently accepted wisdom is included in each of these categories, there is no mention of skills of observation or the ability to challenge, to deal with ambiguity, to consider alternative answers/procedures, or to imagine new ideas. In the detailed explication regarding mathematics and science, knowledge of facts, skills in observation, and analysis are well covered; technology is not, however, and the only mention of inquiry is under laboratory training, i.e., "the ability to distinguish between scientific evidence and personal opinion by inquiry and question."

The Task Force on Education for Economic Growth of the Education Commission of the States, in its June 1983 report, does note that "technological change and global competition make it imperative to equip students in public schools with *skills that go beyond the basics."* In discussing improving academic experience for students neither technology nor innovation is included but the report does note:

> The goal should be both richer substance and greater motivational power: to involve students more enthusiastically in learning, and to encourage mastery of skills beyond the basics—problem-solving, analysis, interpretation, and persuasive writing, for example.

A comprehensive research study by the National Association of Secondary School Principals and the Commission on Educational Issues of the National Association of Independent Schools delivered a preliminary report, *A Study of High Schools,* in 1983. The most prevalent observation is students' docility, lack of engagement with ideas, and disinterest in school. After discussing the need for policy makers to understand the complexities of adolescence, the processes of learning, and the nature of teaching (as well as the variety of the human beings who are teachers!), the report goes on to say, "Secondary schools should be primarily places where young citizens learn to use their minds. One learns to reason—to imagine, to hypothesize, to analyze, to synthesize—by practice." Here is another definition of new educational objectives!

Recommendations for Action: What Can We Do?

Clearly our national need is to set and to implement educational goals, particularly for mathematics, science, and technology, that are more in tune with our times and our students—and with the nature of these fields of inquiry themselves.

CALLS FOR LEADERSHIP AND ONGOING ASSESSMENT

Considering the lack of mechanisms now in place to recommend, implement, and monitor the necessary changes, the NSB commission points to the need for new leadership to set and implement new goals at the national and state level, involving all sectors of American society. Among its forty-one recommendations was that a "National Education Council" be established, as well as state councils (such as those already initiated in several states) to provide focus, coordination, direction, and midcourse corrections, as well as assessment of student participation and performance, in order to achieve constructive change for education in mathematics, science, and technology. Another pertinent recommendation was that local school boards take appropriate steps to form partnerships with institutions and individuals who can aid in science and technology education.

EARLY EXPOSURE: EARLY ADVANTAGE

It is important to recognize, *particularly* in the mathematics and sciences, that early talent identification means early advantage. Early advantage compounds during the student's subsequent life and creates an increasing separation between the talented student's achievement and that of others.

As in the critical work of Robert Merton, American science can be demonstrated to be meritocratic, in that identified talent tends to be rewarded on the basis of performance rather than origin. The ultraelite continue to come largely from the middle and upper classes.

In her definitive study of U.S. Nobel laureates, Harriet Zuckerman identifies early advantage and early identification as common to all laureates interviewed. The proportion of all scientists coming from families of skilled and unskilled workers has, over the past fifty years, trended slightly upward, but no change is evident in the social origins of laureates. Thus, although inequalities in the socioeconomic origins of American scientists at large have been reduced during the past half-century, this has not been true for the ultraelite in science. According to Zuckerman, the origins of Nobel laureates remain highly concentrated in families that can provide their offspring a head start in system-recognized opportunities.

To quote Lilli Hornig, in a 1982 National Research Council annual review of issues:

> The traditional approach to science teaching is grounded in a belief that quantitative talent appears early in life, and that potential scientists are likely to be identified in adolescence or sooner. Their education is therefore designed on a largely sequential model that leaves little room for late bloomers or those whose social condition made early identification and fostering of talent unlikely. The model persists through higher education and scientific careers: those who move fastest are likely to be labeled "best."

Giftedness in the socioeconomically disadvantaged child may be discouraged rather than encouraged in getting on an upward track of achievement. The work of Donald McKinnon documents a commonly held impression that gifted children may not be valued by teachers in the classroom or parents at home because they are not necessarily held to be good prospects to succeed as adults; children themselves do not necessarily want to be gifted in the home/school value context of giftedness.

One also recalls here the overriding influence of the "Hawthorne effect": student self-image and the teacher's image of that student are of overwhelming influence in teaching and learning.

The seminal work of Seymour Papert and his students and colleagues at M.I.T. helps to update these speculations and observations about early identification and advantage. His book, *Mindstorms*, notes the deeply embedded assumption in our culture that the appreciation of mathematical beauty and the experience of mathematical pleasure are accessible only to a minority, perhaps a very small minority, i.e., uniquely creative, gifted, and talented children. French mathematician Henri Poincaré accepted this as a truism while expostulating that the distinguishing feature of the mathematical mind is not logic but aesthetics and that this aesthetic sense is innate. Papert suggests that, if Poincaré's model of mathematical thinking (i.e., predominantly aesthetic, not logical) is correct, then the affective and aesthetic dimensions of mathematics should be included in curricula. If this were done, could hidden talent thereby be unlocked and the assumption that the mathematical mind's aesthetic sense is entirely innate be challenged?

The NSB commission found "a striking relationship between achievement in mathematics, science, and technology and the early exposure of students to stimulating teaching, good learning habits in these fields, and enrichment by regular exposure to informal educational activities." It thus recommends that top priority should be placed on providing increased and more effective instruction in mathematics, science, and technology in kindergarten through grade 6 (K–6). From the foregoing discussion it seems clear that it is essential to introduce mathematics in the earliest grades (K–3) in ways that encourage and develop innate imagination, gamesmanship, and numerical and spatial sense. Thus our society might involve a broader cultural grouping than predominantly upper- and middle-class white males in an education leading to innovative work in science and/or technology. To enable students' minds to be opened and remain open to mathematics and sciences in high grades requires the type of creative and comprehensive early introduction recommended by Papert.

IMPROVED INSTRUCTION: CONTINUING ADVANTAGE

Discrepancies observed between recommended educational objectives and the learning that goes on in most classrooms, as summarized above, indicate the quantity and quality of improvement that must be made in instruction. To the multitude of challenges and obstacles faced by education in general, the sciences add the exploding complexity of the subject itself and the critical need for laboratory and/or field experience for its understanding. In analyzing why the extensive involvement of science teachers in National Science Foundation sponsored teacher institutes in the 1960s and 1970s did not lead to more long-lasting improvement in the classroom, Arnold Arons, Emeritus Professor of Physics at the University of Washington, notes the failure of most teacher institutes to guide teachers slowly and carefully through the intellectual experiences they were subsequently to convey in their classrooms: "Most teachers had developed little genuine understanding of scientific subject matter in their previous school and college courses and were very nearly at the same level of conceptual development as their pupils." The interdependence of school and college teaching is clear. He also notes that science teachers, particularly in elementary and junior high school, need much better logistic support (in time, equipment, and materials) if they are to successfully try new curricular and teaching concepts.

Nevertheless, the federally funded efforts in curriculum development and in teacher training between 1957 and 1977 yielded a sound base of experience, partnerships between teachers, scientists, mathematicians and professional societies on which to build for the future. High school curricula were developed that could be readily updated and that are good preparatory sequences for those preparing for advanced careers in the sciences. New instructional strategies were explored with model materials to support them. Most importantly, new views of science education, along the lines of the objectives discussed above, were promulgated, which include philosophical, historical, sociological, technological, and humanistic dimensions. Thus with renewed commitment, agreement on new educational objectives (including technology), and broad-based leadership, a reservoir of knowledge and experience is available to embark on a massive effort to retrain and to improve the continuing education of the nation's mathematics and science teachers.

The NSB commission suggests such a commitment at the national level, in partnership with states, to accomplish extensive retraining of the nation's elementary teachers and secondary mathematics and science teachers over a five-year period. Coupled with these recommendations are others relating to increasing graduation requirements and time spent on mathematics and science for high school students, as well as for baccalaureate degrees for future teachers. More mathematics and science and technology for everyone is the consensus today.

To ensure that the objectives of teacher training are geared toward the approaches in mathematics, science, and technology recommended above will require a fundamental change in attitude about these subjects, and about teaching, by administrators, parents, and community leaders. Citizens and educators must look at what is really happening in the classroom and beyond. It is hoped that their tremendous current interest and focus will prompt those concerned to take the studies by the mathematics, science, and engineering societies very seriously and press these recommendations widely.

Needed change in the role and function of the teacher may develop through computer science education. With both teacher *and* student relatively new to this field, it is legitimate for them to learn together. The role of the teacher, in these classes, is more that of an information transfer agent, the person who knows and has access to learning resources. In the best cases, the teacher understands the students and can guide, manage, and facilitate the transfer of knowledge and understanding. Teachers thus assume their proper role—the agent for learning—not the source of all knowledge.

Leadership of future technological innovation in the nation requires young people capable of both subject matter mastery and innovative thought, so ways must be found to help communities and school administrators understand that student behavior reflecting both capacities is in line with new educational objectives and is to be rewarded. Teacher training must incorporate such recognition, as must administrative procedures in schools and in teachers' continuing education.

Gifted and talented students can block out learning if there is too much drill and repetitive practice, even though others may need it. How can we protect the learning needs of all students; move all to a certain degree of general, as well as technological and scientific literacy; and, at the same time, deal with the evident and not so evident talent that needs the opportunity to develop inventive thinking? The only answer is, of course, improved training, continuing education, and improved resources for teachers.

NEW LEARNING ABOUT HOW PEOPLE LEARN

Among the new resources available for direct assistance to teachers and to curriculum developers to help improve the effectiveness of teaching and the efficiency of learning is the promise of current research in the cognitive and behavioral sciences. Perhaps this field will supply ways to address the special needs of the gifted student, the "breakaway mind," and the underachiever. Publications by the National Science Teachers Association (e.g., *What Research Says to the Science Teacher*) and recommendations of the various commissions and studies cited earlier certainly indicate a general and growing recognition among professionals of what this field has to offer them. For example, theoretical representations of knowledge structures needed to represent problems in physics, elementary arithmetic, and electronics have been developed in the form of schematic mental models, from which instructional methods are being designed to increase students' skills in representing problem information. Current analyses are showing how understanding of general concepts can facilitate learning and performance of correct procedures, as well as understanding of the procedures themselves.

Some of the most interesting work, which has immediate application to the classroom, is in the domain of physics—the subject so often perceived to be the "hardest" and most "beyond" the average pupil. Students apparently bring to the classroom significant misconceptions of general principles which persist despite their instruction, so that their subjective or qualitative understanding of the principles of physics (i.e., gravitational force and laws of motion) is inconsistent with and *thus interferes with* the formulas thrust on them in the classroom. Instructional methods that take into account students' prior conceptualizations, especially instructional materials that use the capabilities of computer simulation to represent systems that behave according to ideal principles, can help students make great strides in comprehension. This is an example of applying computer simulation in order to improve understanding of fundamental principles—the aspect of learning previously considered the exclusive domain of the gifted or scientifically and mathematically intuitive student. How much greater a percentage of the minds of our student population can be opened to such understanding and new potential in innovative work remains for the next generation of teachers —those using computer simulation together with other improved instructional techniques—to determine.

TECHNOLOGICAL AIDS: COMPUTERS

Undoubtedly the most significant resources now available to help teachers increase their effectiveness and expand and increase student learning are the technologies: computers, educational television, videotext data bases and computer based telecommunications, videodiscs and intelligent videodiscs, and robotics.

Computers are the most widely considered technology in the current educational scene and can be used in three distinct modes: learning *about* computers, the most widely used application in schools to date; learning *with* computers (i.e., drill-and-practice and tutorial), the most widely researched area; and learning *through* computers, the area with the most exciting potential for future computer impact on learning. Student and teacher use of computers as aids to learning and teaching is growing and developing at an exponential rate as the cost of hardware decreases and the variety and capacity of hardware increases. There is already strong evidence that computers, used in the "learning-through" mode, make significant contributions to the learning experiences of children in a variety of disciplines (experience with the LOGO language of Papert is an example). Even though there is much less evidence in other areas of application, computers used in the "learning-about" and "learning-with" modes have a great deal to offer educators and students as well.

As explicated earlier, one can anticipate that student interactive work with computers will be literally "mind expanding" as courseware is developed through partnerships between learning psychologists, artificial intelligence specialists, teachers, and subject matter specialists. The potential of students educated through such modes to contribute, in their turn, to innovative work in technology can only be prognosticated, but it should be extraordinary.

As with any change and with any new technology, there are certainly problems involved with incorporating computers most effectively in our educational system.

1. The overall quality of existing courseware is very low.
2. Since there is a well-articulated consensus that *all* teachers should be computer literate and that mathematics and science teachers should have special facility for using computers as aids in instruction, training all teachers to develop and maintain such skills will require a monumental investment in time, talent, and money.
3. The investment cost estimate to develop an adequate base of quality computer courseware in mathematics, science, and technology for all the nation's schools, K–12, is in the hundreds of millions of dollars, and this work should proceed in tandem with curriculum development.
4. Although several fine efforts have been developed on a small scale,

ongoing review of existing courseware and dissemination of results and recommendations require a capacity not now present on the national or regional level.

5. There are serious inequalities of computer access and computer instruction between those schools and school districts that are privileged or targeted and those that are not. The danger is that computer instruction and access will be another case of the "rich getting richer." In this case the "rich" will become those with a particular potential for innovations in this technology. Another consequence of inequality of access is that computers tend to be used more for remedial work in socioeconomically disadvantaged schools than for more creative and advanced use. Computers for drill and practice do not put the child in control or give the child the sense that the child can master the computer rather than vice versa—learning *through* computers does.

An important development in learning with and through computers is already, and will increasingly be, through informal learning environments. These have several advantages over schools, including access by everyone in the community and creation of a nonjudgmental climate without the time constraints of more formal environments. As technology becomes incorporated in school programs, "hands-on" experience with technology may have to depend on such out-of-school access as technology centers and participating science museums. Examples include the Capital Children's Museum in Washington, community based centers like Playing to Win in New York, and ComputerTown USA in Menlo Park, California. However, it is the private home that may be the most powerful influence of all.

Educators must develop ways to take advantage of home computers and to build cooperative relationships with parents in acquisition of hardware and courseware. Some school systems are already involving parents in computer education and enabling school computers to go home in an effort to redress the economic barrier to home computers for many families. School structures and classroom design will undoubtedly and beneficially be forced to change as this technology becomes more widely distributed.

With respect to other technologies, the impact of educational television is strongly favorable, especially when it is accompanied by support documentation for teachers and students. Examples on public television for the precollege and general adult population are the "Nova" series, "3-2-1 Contact," "Connections," and the extensive use of television by colleges, particularly for distance learning. The combination of telecommunications with computers is also one that can be expected to be particularly fruitful for distance learning, as well as learning in rural areas. The other technologies mentioned are so new to education that the only evidence of effectiveness is anecdotal—although positive.

These new technologies have unique potential in expanding the capacity of teachers, particularly in relation to education about the technologies themselves, as well as to excite interest and achievement in mathematics and science where the nonjudgmental, discovery-learning environment provided by computers, television, and related technologies has a special value. Partnerships of industry, higher education, and schools are essential to set guidelines and develop strategies ensuring the most effective introduction of these technologies and the most effective courseware. Leaders in technology, government, business, and higher education would be well advised to give these issues top priority, as the NSB commission recommends.

PARTNERSHIP AMONG SCHOOLS, HIGHER EDUCATION,
AND INDUSTRY

Such partnerships have high potential to increase the effectiveness of schools and colleges by providing better market information and community resources. Considering the direct interest, indeed, proprietary interest, that higher education has in secondary schools—either in their product, i.e., students who will purchase university services, or in their market, i.e., teaching positions for university graduates—it is surprising how independently each has functioned in recent decades. Neither has ordinarily been in the position, on a systematized or mandated (let alone recommended) basis to give the other useful feedback concerning the efficacy of its product—the teachers trained or the students enrolled. With new economic constraints on universities and marked dissatisfaction with schools, there is increasing talk about and beneficial development of such partnerships. One hears of more and more initiative on the part of colleges and universities to share information on student participation and performance with the sending high school, particularly with respect to mathematical and writing proficiencies. Can schools find the same capacity to feed back information to colleges on the efficacy of their preparation of teachers?

Will students' capacity to be innovative and/or productive in technological fields become part of this feedback to schools? This possibility can occur only if the value of such student qualities is identified and given emphasis. If this message could then come from colleges to schools, a direct effect in developing teaching methodologies encouraging such thought and action could be anticipated.

Increasing interaction of colleges and neighboring schools in student mentor projects, summer teaching assistantships for teachers, and adjunct teacher programs has already encouraged, and will further increase, informal feedback, and therefore more systematic communication. Programs to encourage student work in laboratories, such as those initiated by the New York Academy of Sciences, the Work in Technology and Science Program at M.I.T., the North Carolina School of Science and Mathematics and Duke University Medical Center, the University of Michigan, and the University of Rochester, can be cited as among those that were initiated tentatively and then more fully developed as success bred success in the relations between high school students and college mentors and those between high school teachers and college laboratories and professors.

Partnerships between schools and industry appear to be burgeoning, as are those between schools and higher education. A review by educator Michael Timpane indicates that the business community thought and acted as though it "owned" the schools until twenty or twenty-five years ago. Business and professional men dominated school board membership and were, until the 1950s, the most powerful and consistent source of civic support for the schools. School administrators increasingly identified themselves as managers in the image of the American businessman. With radical changes in the political economy of urban schools in the mid- to late 1960s, the corporate representatives gradually left the scene, particularly in urban areas. The current image of schools as viewed by corporate officers tends to be one set in rather absolute terms such as "declining test scores, unruly students, unworkable innovations, and newly militant teachers." Corporate officials have had little personal contact to soften this stereotype.

Corporations have recently begun to invest substantial financial and political resources in the rejuvenation of city schools. Many programs have started: Adopt-a-School and Join-a-School programs and privately supported foundations to assist public schools. Donated equipment, loaned adjunct teachers, and summer employment for students and, increasingly, for teachers are among the forms of participation initiated by science and technology based industries. Such programs show a unique potential for providing laboratory experience and up-to-date equipment information for teachers and students.

With respect to innovation in teaching and learning, one could suggest a very special role for higher education and industry for students with high potential in this area. Volunteer scientists and engineers in the schools or in their own laboratories can encourage students with the innovative risk-taking type of thinking that is so hard to provide in large classes and understaffed schools. In out-of-school settings, other types of thought and action can be explored that might, in a school setting, threaten other students and personnel who do not have the confidence or perception to deal with them.

MUSEUMS AND TECHNOLOGY CENTERS

Museums and technology centers have been leaders in demonstrating how much more readily out-of-school settings adapt themselves to encouraging creative and innovative approaches to and learning in science. In recent decades, a new sort of educational institution has become popular around the world—the science and technology center. Although the name "museum" is often included in the title, they are quite different from the traditional museum. Rather, they are populist, interactive facilities designed to expose the inner workings of natural phenomena and man-made processes and are alive with a multitude of participatory exhibits and educational programs. Their origins in Germany, England, and the U.S. (beginning with The Franklin Institute) are closely identified with inventions and technology. Although, for the most part, they have emerged in the United States only within the last fifty years, they raise pertinent questions about places for technology education today.

A historical review of these institutions by V. J. Danilov, director of the Chicago Museum of Science and Industry, makes it clear that for the last three centuries there has been a general assumption that both technology education and stimulation of invention and innovation were best done in a physical setting where things could be observed. Another assumption was that technology education provided through an exhibition was good for everyone, but mathematics and science were for schools and the elite. Is it because of this assumption that technology *is* interesting to everyone that there is virtually no technology education in schools today? Inventions and innovation have not been considered to be part of the school's training responsibility, but rather for out-of-school opportunities.

In the late nineteenth century, G. Browne Good, a director of the U.S. National Museum (operated by the Smithsonian Institution) said:

> The museum of the future must stand side by side with the library and the laboratory, as part of the teaching equipment of the college and the university, and in great cities cooperate with the public library for the enlightenment of the people.

This is, indeed, a role that the NSB commission is recommending for today. With 150 million Americans annually visiting museums, the role such centers, whether large or small, could play in greatly expanding the horizons of science and technology education for students and adults is clear. With their capacity to emphasize innovation and technology in an informal setting, museums and technology centers could serve as unique resources for teacher education, particularly for those teaching at the elementary level.

YOUTH ORGANIZATIONS, INCENTIVES, PUBLIC AWARENESS

Americans have been great "joiners" since the founding of the Republic; yet there is a lack of general recognition of the vast array of opportunities afforded our youth through the over 250 adult sponsored youth organizations that enroll millions of youths in groups, troops, teams, and clubs. A veritable army of adult volunteers and well over 50,000 staff members are involved. Opportunities abound through our youth organizations for developing educational programs, scientific and technological experiences, awards, and incentives to encourage student participation. These informal educational opportunities allow participation of interested professionals in a host of ways.

The positive effect of awards, rewards, and honors on the development of interest and self-confidence in youth is well documented. The Westinghouse Talent Search, science fairs, and various national and international competitions (such as the Mathematics Olympiad and Math-Count) are well-known examples. Such programs serve as incentives to teachers as well as students. With all that has been said earlier of the limitations of schools and college administered written tests of cognitive ability, there are almost unlimited opportunities to develop incentives and rewards for creative, experimental, and innovative activity through the entire informal education system—youth organizations, church groups, museums, and local television stations. While more direct opportunities are clearly available through these channels for affecting attitudes, such programs help raise public awareness and disseminate information about the nature of science and technology in ways that are not restricted by the academic organization of formal educational institutions.

EXEMPLARY PROGRAMS, "MAGNET SCHOOLS"

The magnitude of the tasks ahead is overwhelming: teachers to recruit, train, and retrain; public attitudes to change and expand; new curricula to develop and old curricula to revamp; equipment to evaluate, purchase, and use; and all the students, young and old, to reach. The existence of many wise and wonderful teachers and students in excellent exemplary programs lends not only hope but the expectation of success to this task. Lasting change in education has come from discrete individual steps—the development of programs that work in given communities has a "ripple effect," encouraging others to follow or adapt them to other situations and people. The NSB commission suggests strongly that the development of such exemplary programs, particularly "magnet schools," where appropriate, be given a national priority. This strategy is based upon the recognition that change cannot be made all at once but should be initiated in special situations and then disseminated through outreach, example, and competition.

The "ethos" of a school is key. In some cases model programs can be developed within the school or the school can become a "model." In others, the environment in and around the school is such that change and positive attitudes can only be developed with the creation of a new school or a school-within-a-school where student and teacher commitment to higher goals becomes the admissions credential. The strategy of focusing the school program around a subject area such as the arts, science and mathematics, or engineering has been used successfully in several inner city schools. Again, it is the positive attitude about the subjects chosen for emphasis *and* the assumption that all children can succeed within it that make the difference. Those interested in finding fertile environments for special incentives or special programs to develop student interest in innovation in technology and the sciences might find it easiest as a first step to initiate such a program within the receptive environment of the over 1,100 "magnet" elementary and secondary schools now in operation across the nation.

Summary

This chapter has concentrated its discussion and recommendations on elementary and secondary schooling simply because this is where the quality of future students and workers is controlled and innovation first encouraged. Change and improvement at this level will both force and catalyze improvements in colleges and in the workplace—and vice versa.

Both innovation and technology can and must be included in our objectives for the learning of all students. Our nation has reached the point in its development and maturation where equal opportunity required for and by its citizens must include the early development of the creative, investigative mind as well as skills. This is a right for all and, according to the many resounding calls for action, a need of the nation. The liberal arts, the Jeffersonian, and the Jacksonian traditions for education can no longer be left just to develop side by side; they must be integrated—perhaps to create a recombinant form. To focus on the growth of young minds, both for innovative thinking and for technological understanding, will require also the integration of student centered approaches, particularly those utilizing educational technologies—an update of progressivism, if you will.

The new educational objectives called for include a redefinition of mathematics to meet the needs of all, as well as be of interest to all. Also included is a reinterpretation, both for the public and teachers, of the essence of science and technology that can involve all in searching enthusiastically for understanding and mastery. To include representatives from our total population in our future pool of scientists and engineers, new ways must be found which develop creative and aesthetic awareness and which introduce stimulating approaches to mathematics and science beginning in kindergarten and primary school and, with instruction becoming more comprehensive and rigorous, continuing throughout the later grades and secondary school. Only by early identification and advantageous opportunity can latent talent develop sufficiently to become both interested and productive in the research or industrial laboratory. Only by continuing efforts to involve all students and all community resources in secondary and college mathematics, science, and technology education can all our citizens be prepared for the future and can our leaders in innovation be developed and supported.

To direct our educational approaches toward these goals, a "sea change" is required in the expectations of schools, parents, and professionals regarding teacher training, curriculum development, and community-school relationships and the nature of science and technology themselves. With public awareness of the need for increased and broader emphasis on these subjects throughout school and college and with improved training, resources, and objectives available for teachers, the satisfactions and rewards of the profession should be so vastly improved that their ranks should swell with both quality and quantity. In terms of the long-run security of this nation resulting from the satisfaction and productivity of its citizens, its professional leadership, its political judgment, and its continued capacity for innovation in technology, there can be few other public policy goals of such ultimate consequence.

Index

About *The American Assembly*

The American Assembly was established by Dwight D. Eisenhower at Columbia University in 1950. It holds nonpartisan meetings and publishes authoritative books to illuminate issues of United States policy.

An affiliate of Columbia, with offices in the Sherman Fairchild Center, the Assembly is a national educational institution incorporated in the State of New York.

The Assembly seeks to provide information, stimulate discussion, and evoke independent conclusions in matters of vital public interest.

AMERICAN ASSEMBLY SESSIONS

At least two national programs are initiated each year. Authorities are retained to write background papers presenting essential data and defining the main issues in each subject.

A group of men and women representing a broad range of experience, competence, and American leadership meet for several days to discuss the Assembly topic and consider alternatives for national policy.

All Assemblies follow the same procedure. The background papers are sent to participants in advance of the Assembly. The Assembly meets in small groups for four or five lengthy periods. All groups use the same agenda. At the close of these informal sessions, participants adopt in plenary session a final report of findings and recommendations.

Regional, state, and local Assemblies are held following the national session at Arden House. Assemblies have also been held in England, Switzerland, Malaysia, Canada, the Caribbean, South America, Central America, the Philippines, and Japan. Over one hundred forty institutions have cosponsored one or more Assemblies.

ARDEN HOUSE

Home of The American Assembly and scene of the national sessions is Arden House which was given to Columbia University in 1950 by W. Averell Harriman. E. Roland Harriman joined his brother in contributing toward adaptation of the property for conference purposes. The buildings and surrounding land, known as the Harriman Campus of Columbia University, are 50 miles north of New York City.

Arden House is a distinguished conference center. It is self-supporting and operates throughout the year for use by organizations with educational objectives.

AMERICAN ASSEMBLY BOOKS

The background papers for each Assembly are published in cloth and paperbound editions for use by individuals, libraries, businesses, public agencies, nongovernmental organizations, educational institutions, discussion and service groups. In this way the deliberations of Assembly sessions are continued and extended.

The subjects of Assembly programs to date are:

1951—United States-Western Europe Relationships
1952—Inflation
1953—Economic Security for Americans
1954—The United States' Stake in the United Nations
——The Federal Government Service
1955—United States Agriculture
——The Forty-Eight States
1956—The Representation of the United States Abroad
——The United States and the Far East
1957—International Stability and Progress
——Atoms for Power
1958—The United States and Africa
——United States Monetary Policy
1959—Wages, Prices, Profits, and Productivity
——The United States and Latin America
1960—The Federal Government and Higher Education
——The Secretary of State
——Goals for Americans
1961—Arms Control: Issues for the Public
——Outer Space: Prospects for Man and Society
1962—Automation and Technological Change
——Cultural Affairs and Foreign Relations
1963—The Population Dilemma
——The United States and the Middle East
1964—The United States and Canada
——The Congress and America's Future
1965—The Courts, the Public, and the Law Explosion
——The United States and Japan
1966—State Legislatures in American Politics
——A World of Nuclear Powers?
——The United States and the Philippines
——Challenges to Collective Bargaining
1967—The United States and Eastern Europe
——Ombudsmen for American Government?

1968——Uses of the Seas
——Law in a Changing America
——Overcoming World Hunger
1969——Black Economic Development
——The States and the Urban Crisis
1970——The Health of Americans
——The United States and the Caribbean
1971——The Future of American Transportation
——Public Workers and Public Unions
1972——The Future of Foundations
——Prisoners in America
1973——The Worker and the Job
——Choosing the President
1974——The Good Earth of America
——On Understanding Art Museums
——Global Companies
1975——Law and the American Future
——Women and the American Economy
1976——Nuclear Power Controversy
——Jobs for Americans
——Capital for Productivity and Jobs
1977——The Ethics of Corporate Conduct
——The Performing Arts and American Society
1978——Running the American Corporation
——Race for the Presidency
1979——Energy Conservation and Public Policy
——Disorders in Higher Education
——Youth Employment and Public Policy
1980——The Economy and the President
——The Farm and the City
——Mexico and the United States
1981——The China Factor
——Military Service in the United States
——Ethnic Relations in America
1982——The Future of American Political Parties
——Regrowing the American Economy
1983——Financial Services
——Technological Innovation in the '80s
1984——Alcoholism and Related Problems
——Public Policy for the Arts
——Canada and the United States

Second Editions, Revised:

1962——The United States and the Far East
1963——The United States and Latin America
——The United States and Africa
1964——United States Monetary Policy
1965——The Federal Government Service
——The Representation of the United States Abroad
1968——Cultural Affairs and Foreign Relations
——Outer Space: Prospects for Man and Society
1969——The Population Dilemma
1973——The Congress and America's Future
1975——The United States and Japan